Space
Traveller's
Handbook

Space Traveller's Handbook

Every man's comprehensive manual to space flight

Michael Freeman

HAMLYN
London • New York • Sydney • Toronto

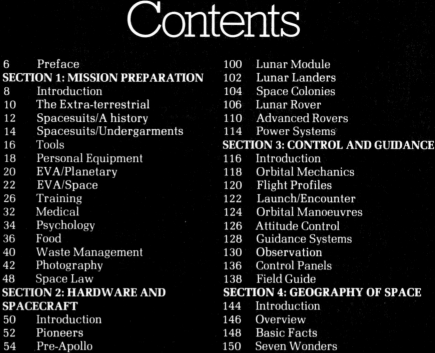

Contents

A QUARTO BOOK
Published by
The Hamlyn Publishing Group Limited
London • New York • Sydney • Toronto
Astronaut House, Feltham, Middlesex,
England
First published 1979

This book was designed and produced by
Quarto Publishing Limited,
32 Kingly Court, London W1, England

Art Director: Robert Morley
Editorial Director: Jim Mallory
Designer: Neville Graham
Editor: Jim Roberts
Illustrations: Bill Easter
Diagrams and Illustrations: David Worth
Diagrams: David Staples
Filmset in Britain by Abettatype Ltd, London
Colour illustrations originated in Britain by
Scankolor, Ilkley
Printed in Hong Kong by Dai Nippon Printing

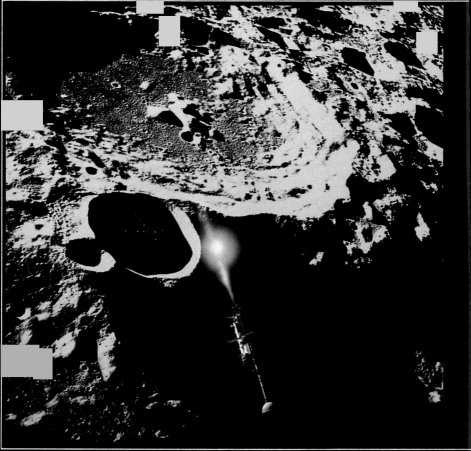

Preface

Exactly 100 years ago, on the 12th of April 1961, Man entered space. Although by now we are accustomed to the rapidity of change in technology and society, it nevertheless seems strange to imagine that most of Man's history was confined to the Earth. The early explorations of Earth Space were undertaken at enormous expense, and had little direct effect on the majority of the human race. Now, however, space is a natural medium for a growing number of people, and the public at large has taken the exploration of the Solar System in its stride.

Surprisingly, until now there has been no comprehensive guide for the prospective traveller in these regions. Some people still approach certain aspects of a space voyage with trepidation. With the publication of this Handbook all doubts and uncertainties may be dispelled. What we hope to present is a full briefing for virtually any spaceflight currently available.

As is the fate of all guides that attempt to stay abreast of modern developments, it is feared that revised editions will be necessary in the coming years. The Editors will therefore gratefully receive any comments that readers may have in the light of their own experiences.

The Editors
London
April 2061

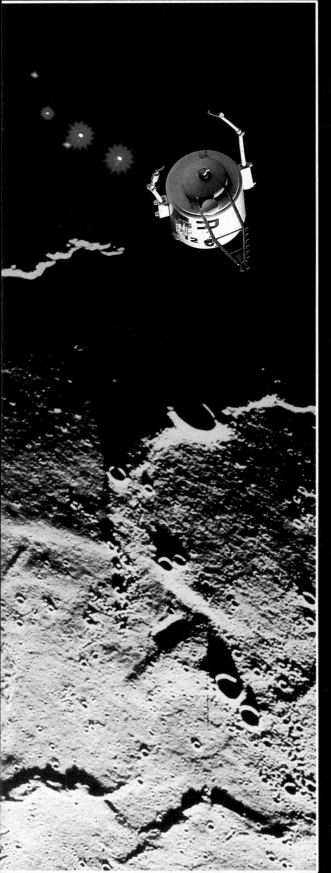

Mission preparation

THE HOSTILITY of *all* extra-terrestrial environments, from deep space to the Martian desert, demands that the prospective space traveller be thoroughly familiar with the conditions that await him (or her) and be well-trained in meeting them.

It is not sufficient merely to have read a few accounts of space travel; a thorough preparation is essential, and indeed is demanded by most carriers. Many training centres exist on Earth as well as in Earth-orbit transfer stations, and it does no harm to be very meticulous in planning a voyage, even if it be no further than to the Lagrange Colonies. In Space, small mistakes can be fatal, and inexperienced astronauts may not live long enough to make large ones.

Having said this, mission preparation need not be a chore, but can be an exciting prelude to the unforgettable experience of space travel.

A spacetug manipulator module manoeuvres over the lunar surface. The lights of a small base can be seen just beyond the terminator.

The Extra-terrestrial Varieties of hostility

EVEN WITHIN the Solar System, the range of environments is extreme when compared with Earth conditions and with the conditions under which an unprotected human being can survive. There is no environment in the Solar System which does not require a full pressure suit, and some, such as the surface of Venus, continue to present problems which make personal protection of the astronaut impossible.

Of course, the range of temperature, pressure, gravity and radiation levels within which a human being can survive for long periods is much smaller than that for temporary survival conditions. For instance, a man can normally survive two minutes of no oxygen pressure if proper recompression procedures are followed immediately. Or, survival for several days is possible at high Earth altitudes, although at more than about 6000 meters certain functions begin to deteriorate and the body cannot continue to maintain normal operation.

The most important parameters for survival in the Solar System are temperature, pressure, gravity and radiation. It is within these that modern spacesuits are designed. Certain special conditions also apply: the high surface temperature and pressure of Venus is a stumbling block to protective suit design that has not yet been overcome, and phenomena such as Martian duststorms can undoubtedly be hazardous.

ET Environments- key
◯− Temperature °C
✕ Pressure in millibars
⌇ Average ionizing radiation dosage
⚲ Gravity (Earth = 1.0)

EARTH
◯− Mean 22 at surface
✕ 1000 at surface
⌇ 0.1 rem per year
⚲ 1.0

MERCURY
◯− 350 to − 170
✕ 10^{-9} at surface
⌇ 10 rem year (cosmic background) + several rems/hour (solar flare)
⚲ 0.37

MARS
○- Mean -23 at surface
✗ 6 at surface
⌇ Moderate
⚹ 0.38

IO
○- −140
✗ Negligible (Io orbits in a toroidal cloud of storm)
⌇ Instantly lethal
⚹ 0.17

VENUS
○- 480 at surface
✗ 90,000 at surface
⌇ Low cosmic and solar radiation, attenuated by the dense atmosphere
⚹ 0.88 Principally CO_2 atmosphere with highly corrosive compounds

PLUTO
○- −233, the coldest planet in the Solar System
✗ None. Only neon, hydrogen and helium can remain unfrozen.
⌇ Same cosmic background as other regions of the solar system.
⚹ Negligible

Spacesuits A history

OVER THE YEARS, there have been many designs and makes of spacesuit, but, with the exception of the new hard-torso suit most of the differences have been in detail rather than in principle. The basic function of all suits is to pressurize the whole body whilst permitting free movement. This was a fundamental problem for early designers, as a flexible material tends to expand to a more-or-less rigid balloon, whilst a hard suit needs joints, which are inevitably weak links. The answer, reached as early as the Apollo programme, was a combination of both a woven fabric convolute and a series of joints.

Specifically, the spacesuit must do the following:–

1. Maintain a constant pressure of around 3 psi
2. Supply and scrub oxygen
3. Maintain temperature within the normal body range
4. Maintain humidity high enough for comfort and low enough to prevent condensation
5. Protect the wearer against UV radiation, normal background ionizing radiation, and micrometeoroid bombardment

The development of 'link-net' fabric — interwoven slippery strands of Teflon that function like chain mail — and low-torque joint bearings has solved the problem of sustaining pressure. At a regulated humidity, oxygen is supplied either from a portable back pack, on-board storage tanks, or a closed ecology system that recycles used air through a greenhouse. Body temperature is regulated by water flowing through a special undergarment, and multi-layered materials with wire mesh provide basic protection against radiation and particles. Some measure of the technological improvements that occurred even in the early days can be gained by comparing the cost of a Gemini suit — $80,000 — with that of an Apollo suit a decade later — $400,000.

The Gemini helmet, (above), was gold-plated to reflect solar heat and had a visor which filtered out solar radiation.

The standard pressure suit, based on the Hamilton Standard and International Latex Corp. designs of the 1960s and 1970s, is widely available. Reconditioned suits can be purchased at a much reduced price, but prospective buyers are advised to examine the accompanying ISA Spaceworthiness Certificate.

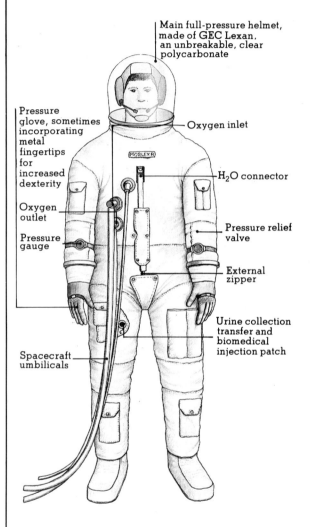

Main full-pressure helmet, made of GEC Lexan, an unbreakable, clear polycarbonate

Oxygen inlet

Pressure glove, sometimes incorporating metal fingertips for increased dexterity

H_2O connector

Oxygen outlet

Pressure relief valve

Pressure gauge

External zipper

Urine collection transfer and biomedical injection patch

Spacecraft umbilicals

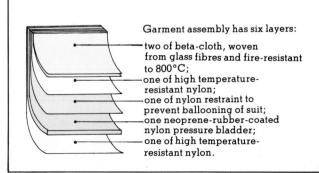

Garment assembly has six layers:

two of beta-cloth, woven from glass fibres and fire-resistant to 800°C;

one of high temperature-resistant nylon;

one of nylon restraint to prevent ballooning of suit;

one neoprene-rubber-coated nylon pressure bladder;

one of high temperature-resistant nylon.

A United States Air Force suit built in 1940, probably by the B.F. Goodrich company. It incorporated accordion bellow joints.

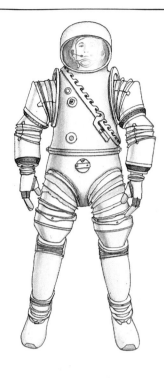

The RX-1 hardsuit: an early, cumbersome hardsuit design which was only used experimentally.

The Mercury "silver" space suit in fact had an aluminized coating to reflect heat. Mobility was rather restricted.

The Gemini suits dropped the aluminized coating, and incorporated umbilical attachment facilities.

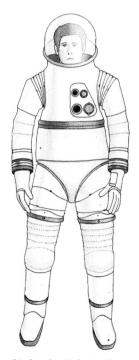

This hard suit design has an integrated helmet and stove pipe joints which make use of rotary seals.

The hard torso suit is a hybrid combining soft fabric arms and legs with a hard-shell torso.

Spacesuits Undergarments/planetary

THE PLANETARY EVA SUIT, essentially the same as the original Apollo lunar suit (known as the Extravehicular Mobility Unit — EMU), differs from the standard pressure suit only in two important respects. An additional but integrated garment gives tougher protection against physical hazards, and a portable back pack handles oxygen, temperature and communications.

Liquid cooled and constant-wear garments

The liquid cooling garment (LCG) is a temporary one-piece underwear, donned prior to EVA. It incorporates a water-cooling system and a urine collection garment. Originally termed 'constant wear garments' (CWG's), intravehicular suits are no longer standardized, and many astronauts use light sportswear or ordinary coveralls. The main criterion is that the CWG should be comfortable enough for prolonged wear and allow unrestricted movement.

The EVA suit (above) allows the astronaut to work on the extravehicular Skylab
Left: astronaut Vance Brand in charge of the command module during the Apollo-Soyuz TP. His proximity to the instrument panel gives a fairly accurate idea of the distinct lack of space which the early astronauts had to cope with. Television and experimental demands added to the problem.

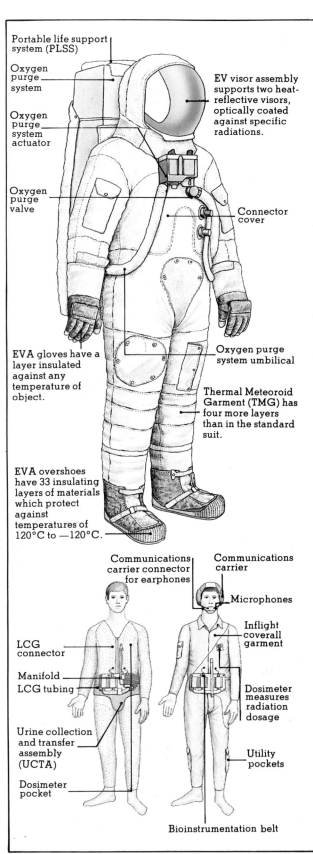

Portable life support system (PLSS)

Oxygen purge system

Oxygen purge system actuator

Oxygen purge valve

EV visor assembly supports two heat-reflective visors, optically coated against specific radiations.

Connector cover

Oxygen purge system umbilical

EVA gloves have a layer insulated against any temperature of object.

Thermal Meteoroid Garment (TMG) has four more layers than in the standard suit.

EVA overshoes have 33 insulating layers of materials which protect against temperatures of 120°C to —120°C.

Communications carrier connector for earphones

Communications carrier

Microphones

Inflight coverall garment

LCG connector

Manifold
LCG tubing

Dosimeter measures radiation dosage

Urine collection and transfer assembly (UCTA)

Dosimeter pocket

Utility pockets

Bioinstrumentation belt

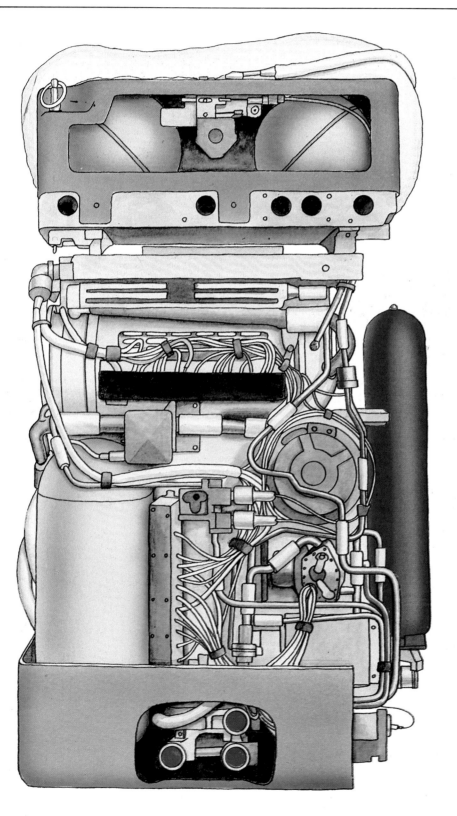

Portable Life Support System (PLSS) as used by Apollo astronauts.
The system contains an activated charcoal bed, oxygen purge system etc. for atmosphere, a pressurizer, insulator and refrigeration unit for maintenance of necessaryt environment, and communications and receiving equipment. Its power source is a silver-zinc battery, with fibreglass cover.

Tools A selection

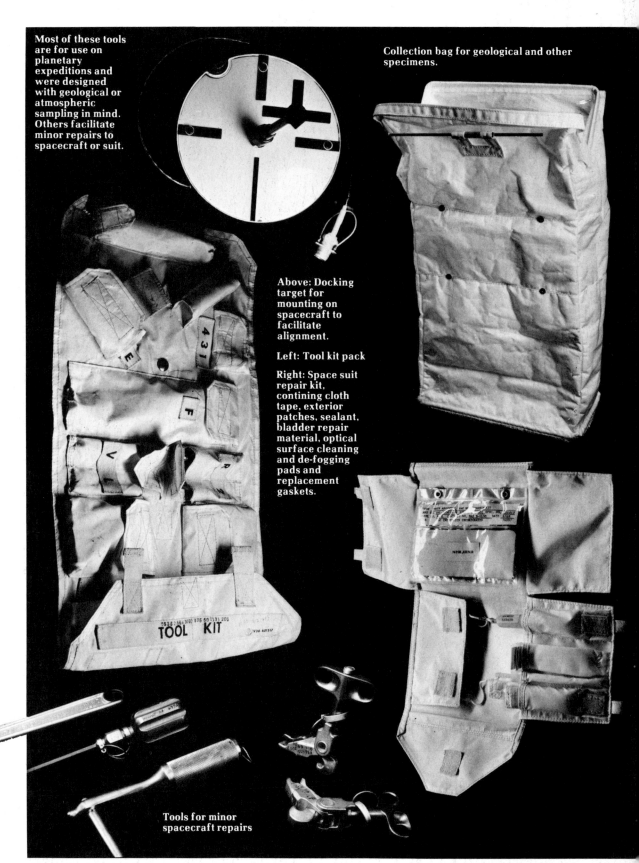

Most of these tools are for use on planetary expeditions and were designed with geological or atmospheric sampling in mind. Others facilitate minor repairs to spacecraft or suit.

Collection bag for geological and other specimens.

Above: Docking target for mounting on spacecraft to facilitate alignment.

Left: Tool kit pack

Right: Space suit repair kit, contining cloth tape, exterior patches, sealant, bladder repair material, optical surface cleaning and de-fogging pads and replacement gaskets.

TOOL KIT

Tools for minor spacecraft repairs

Core sampler

Combined lens and specimen brush for examining specimens in the field.

Gas analysis sample containers for collecting and preserving atmospheric samples. These containers were first used near the lunar surface to collect the rare gases there.

Sample brush for cleaning dust and soil from rock specimens.

Sample bag with folding handle. This was used immediately the Apollo astronauts set foot on the Moon, to ensure that a sample would be returned to Earth.

Soil and rock scoop

Digging and trenching tool

Personal equipment Some necessities

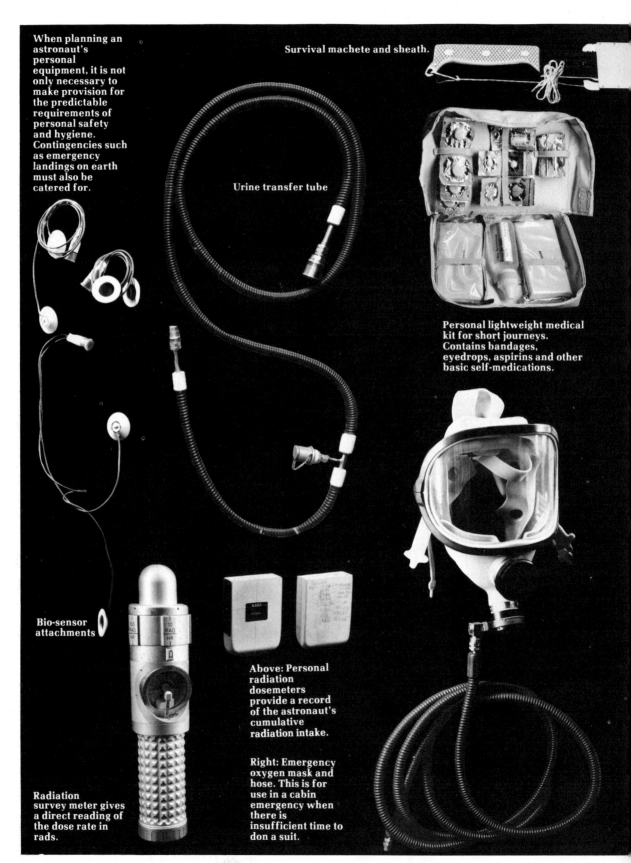

When planning an astronaut's personal equipment, it is not only necessary to make provision for the predictable requirements of personal safety and hygiene. Contingencies such as emergency landings on earth must also be catered for.

Survival machete and sheath.

Urine transfer tube

Personal lightweight medical kit for short journeys. Contains bandages, eyedrops, aspirins and other basic self-medications.

Bio-sensor attachments

Above: Personal radiation dosemeters provide a record of the astronaut's cumulative radiation intake.

Right: Emergency oxygen mask and hose. This is for use in a cabin emergency when there is insufficient time to don a suit.

Radiation survey meter gives a direct reading of the dose rate in rads.

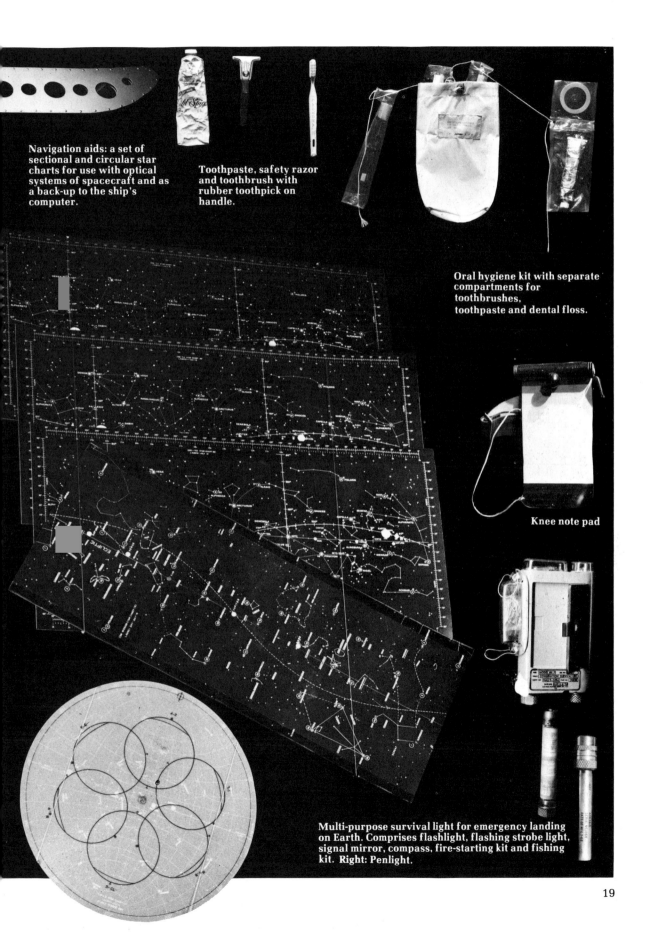

Navigation aids: a set of
sectional and circular star
charts for use with optical
systems of spacecraft and as
a back-up to the ship's
computer.

Toothpaste, safety razor
and toothbrush with
rubber toothpick on
handle.

Oral hygiene kit with separate
compartments for
toothbrushes,
toothpaste and dental floss.

Knee note pad

Multi-purpose survival light for emergency landing
on Earth. Comprises flashlight, flashing strobe light,
signal mirror, compass, fire-starting kit and fishing
kit. Right: Penlight.

EVA Planetary

SURFACE EXCURSIONS must cater firstly for physical environments which can catch even the best prepared traveller unawares. Many known environments are classified into two groups, 'lunar' and 'snowball'. The rest must be treated individually.

The 'lunar' planets and satellites, as the name implies, have surface conditions that resemble the Earth's moon — that is, virtually no atmosphere, and heavily cratered dust-covered terrain that is frequently rocky and occasionally mountainous. They include our Moon, Mercury, Ceres, Callisto, Japesus and many smaller bodies.

The 'snowball' satellites are found beyond the Inner System, are also virtually without atmosphere, and are typically covered in hard frozen methane that forms a low, hummocky terrain. The surface is powdery and snow-like to smooth and ice-like. It is dangerous not only because it is frequently slippery, but because the touch of anything warm, such as a poorly-insulated fuel cell, will flash-vaporize the methane instantly, with unpredictable results. Among the 'snowball' planets and satellites are Pluto, Triton and Umbriel.

All the other bodies in the Solar System have their own peculiarities, as the Geography of Space section illustrates. From a practical point of view, Mars, the most frequently visited planet, is the most important. Being largely similar to the Southwestern deserts of the United States, its conditions can be experienced quite satisfactorily at the Mars Familiarization Center near Phoenix, Arizona. Sand-walking techniques are taught, and a powerful wind-machine array allows dust-storm survival practice.

For *all* planetary excursions:

Train for Low-G

Quite different from both normal Earth-gravity and zero-gravity, low-G activities often fool experienced astronauts who imagine that Space EVA will have prepared them adequately. Locomotive coordination suffers most and there is a general 'slow-motion' feel to most actions. A certain amount of planning is necessary even for simple movements. In walking, the astronaut must deliberately lean forward in order to keep his centre of gravity *over* his feet; if not, the upper part of his body will not keep up with the lower part, and after a few steps he will fall flat on his back. Fortunately, it is not difficult to simulate low-G conditions on Earth with a horizontal walking machine, and all the GEO transfer stations maintain a rotating area close to their hubs at a constant 0.25G.

Guard against suit rupture

The danger of stumbling, due to poor coordination and difficult terrain, lies in rupturing the pressure suit. This would result in explosive decompression (discussed under Medical Health), or in anoxia. From the earliest days, a protective outer suit has been standard for planetary EVA; this gives extra protection against physical hazards and is equipped with re-inforced over-shoes (see p. 14). Nevertheless, although it can withstand the normal range of knocks and scuffs, it is not indestructible. Volcanic rocks, such as are common on Mars, are sharp, and can cause a jagged tear.

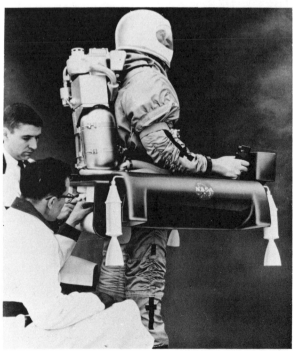

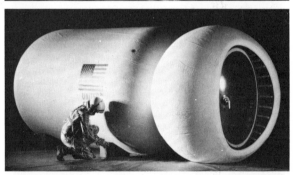

Left: a small personal rocket outfit, with a range which makes it ideal for jaunts around the base. A geologist (above) examines a split boulder at Taurus Littrow. An inflatable shelter (right) for emergencies. It can support lunar travellers for up to two weeks.

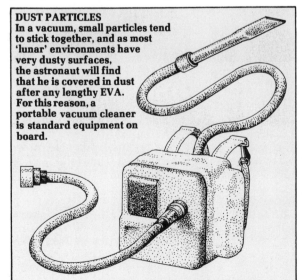

DUST PARTICLES
In a vacuum, small particles tend to stick together, and as most 'lunar' environments have very dusty surfaces, the astronaut will find that he is covered in dust after any lengthy EVA. For this reason, a portable vacuum cleaner is standard equipment on board.

EVASpace(1)

MOST OF THE HARD 'blue-collar' work undertaken in Space involves Extra-Vehicular Activity (EVA). Inherently, EVA poses difficulties for the untrained astronaut, and certain basic techniques must be learnt; however, nowadays the major preoccupation is with operational activity — repair work and construction work, for example.

Tumbling
EVA in Space is never undertaken without thorough training on Earth (see p.26–31). As far as personal movement is concerned, tumbling is a major hazard, easy to initiate with a sudden jerk but difficult to correct. Hand-held propulsion units use high-pressure cold gas for thrust, but a common error is over-correction, either of forward movement or tumbling.

Vision
In most situations in Space, unless in low orbit over a highly reflective planetary surface, there is an enormous contrast between sunlight and shadow. The helmet visors protect astronauts from the Sun's direct rays, but care must be taken with unexpected reflections from bright surfaces. A powerful torch is normally carried for work in shadow.

Restraint
Most tasks necessitate attaching the astronaut in some manner to the surface of the spacecraft. Without a firm restraint, he would fly off into Space as soon as he applied any force to what he was working on. Various methods are used, including foot restraints, variable flexibility tethers and electro-adhesive pads.

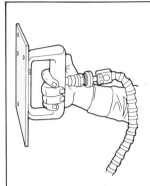

The electro-adhesive pad is a hand-held device that will adhere strongly to any metal surface by means of passing a small current through two metal electrodes. This principle can be used in astronauts' shoes, but this more flexible hand-held method is generally preferred.

Dutch-shoe foot restraints are a simple option common on regular construction sites. Normally used with a waist restraint consisting of belt and tether.

The variable flexibility tether (right) consists of a 3 meter long series of ball and socket links encased in an outer soft covering. A ratchet on the astronaut's belt enables him to apply tension to the links and lock them in any position. Skylab 3 crewmen (facing page) load film in the ATM secured meanwhile by the variable flexibility tether.

A hand-held manoeuvring device stores 0.3 kg of O_2, under pressure in two cylinders. Two triggers release the gas in one of two directions with a maximum thrust of about 1 kg. This allows changes in velocity of about 2 meters/sec. Gemini missions used this model (see above).

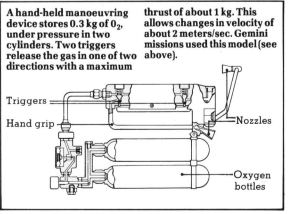

Triggers

Hand grip

Nozzles

Oxygen bottles

EVA Space (2)

Tools for Space

Zero-reaction tools are essential in Space, even when the astronaut is tethered. Any torque produced by a tool would be transmitted to the astronaut, who would either start to spin himself or fly away from the spacecraft. Modern tools, therefore, either employ systems that have no reaction, or compensate internally for reaction. The zero-torque power tool illustrated counter-rotates. Portable space welders, on the other hand, produce no reaction. They come as either Annular Tools or Linear Tools.

EVA Modes

There are three modes of EVA: *surface* EVA is the most elementary, where the astronaut remains on the spacecraft's surface. In *tethered* EVA, the astronaut manoeuvres himself in space but remains attached to the ship and its Life Support System. Finally, for *untethered* EVA, the astronaut must use an independent propulsion system and have his own LSS.

The Manned Maneuverability Unit (MMU)

There are a number of designs in use, but all of these personalized propulsion systems have basically similar characteristics. Using nitrogen gas as a propellant, the typical MMU can operate for up to six hours with normal use. Each of the thrusters can deliver a half-kilo of thrust and can be fine-controlled to give excellent manoeuvring and station-keeping. The unit is flown exactly like a spacecraft.

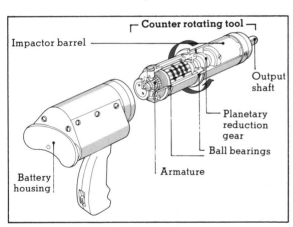

An essential part of any space tool kit is a zero reaction tool like this wrench attachment with motor-handle (above). The attachment eliminates the twisting force or 'torque' which would be experienced with an ordinary power wrench. The tool box (above right) contains other zero reaction attachments (a saw and a drill) along with screwdriver, hammer, adhesive bonding 'astronaut anchor' system, a battery power supply and many other tools.

This box is an early prototype designed by the Martin company.
It is interesting to speculate on the benefits which could accrue to a company whose designs were used in a space programme. They were possibly substantial in terms of pride and prestige, if not in direct financial terms. Competition for the best designs must at times have been fierce.
Speculation on where, when and why private enterprise was used is an interesting topic beyond the scope of this handbook.

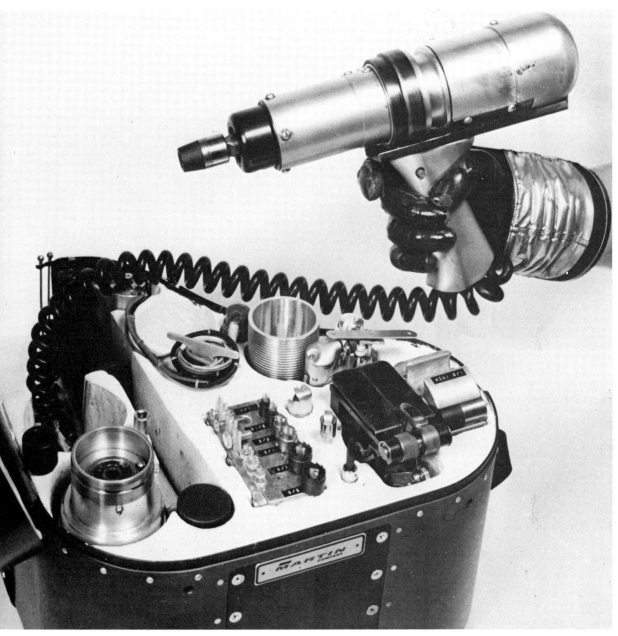

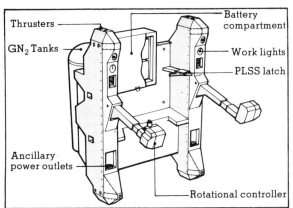

Thrusters

Battery compartment

GN_2 Tanks

Work lights

PLSS latch

Ancillary power outlets

Rotational controller

A Shuttle crew member (far left) equipped with an MMU, approaches a satellite for inspection. The hand controlled device enables him to shuttle back and forth between an orbiter and a spacecraft. Left: a diagrammatic construction of the MMU. The M509 (right) was a jet-powered back-pack similar in design to the MMU. It is seen here being tested for use around Skylab 4. The units are flown very much like a spacecraft and astronauts need no training to speak of.

Training Simulators(1)

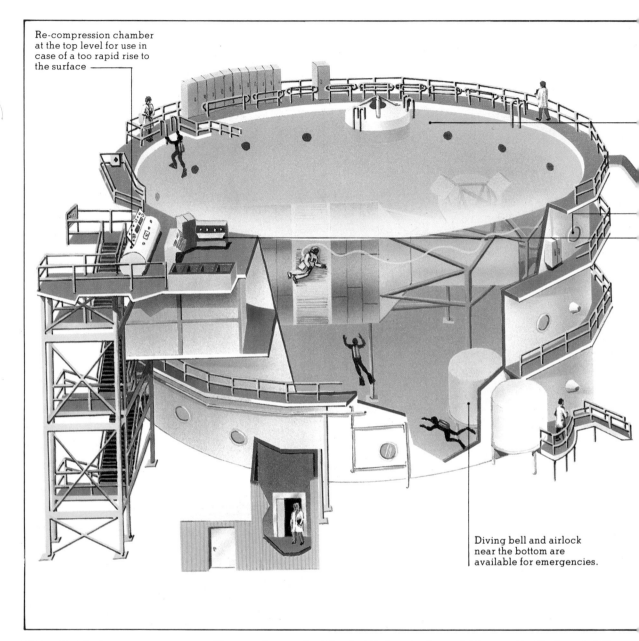

Re-compression chamber at the top level for use in case of a too rapid rise to the surface

Diving bell and airlock near the bottom are available for emergencies.

NOW THAT SHUTTLE LINES provide regular and frequent services to space stations with refined passenger accommodation, astronaut training is not as rigorous as it was, except for the longer missions. Naturally, even shuttle travellers must undergo an approved medical examination, but the standards are quite moderate. It is interesting to look at the *original* requirements set by NASA in 1959 for potential astronauts:

1 Less than 40 years of age
2 Less than 5ft 11in tall
3 Excellent physical condition
4 Bachelor's degree or equivalent in engineering
5 Qualified jet pilot
6 Graduate of test pilot school
7 At least 1500 hours of flying time

The shuttle terminals would be practically deserted if these standards obtained today.

Nevertheless, beyond a simple shuttle journey, some form of training is necessary, and is related to the mission. The basic training systems that are directly related to spaceflight (ordinary physical fitness and health is something that space travellers are deemed capable of maintaining by themselves) are built around simulation. Simulators fall into four groups: zero-gravity simulators, reduced-gravity simulators, high-gravity simulators and disorientation simulators.

The Neutral Buoyancy Simulator — zero-G

Developed in the late 1960 s at NASA's Marshall Space Flight Center in Alabama, underwater simulation of zero-G conditions has now become part

25 meters in diameter by 13 meters deep, the NBS contains 1,400,000 gallons (5,299,574 liters) of water maintained at a constant 30°C, and kept cleaner than drinking water by a pumping-filtering unit and automatic chlorinator.

72 portholes are let into the circumference of the 2.5cm-thick steel plate construction. 36 of the portholes contain high-intensity lamps for illumination.

7 underwater television cameras are controlled from a central panel, with video recorders available for recording activities.

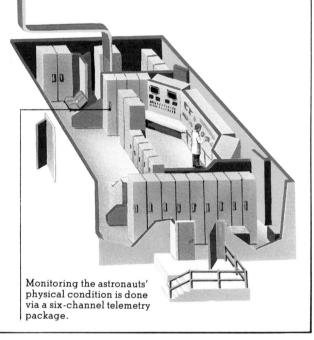

Monitoring the astronauts' physical condition is done via a six-channel telemetry package.

and parcel of space training. It is still the best available simulator for long periods.

After initial experiments with a small water tank, the large Neutral Buoyancy Simulator (NBS) had a crucial role to play in the Skylab programme. A complete Skylab cluster was submerged in the 1,400,000 gallon (5,300,000 liters) tank, to enable astronauts to train at various tasks. Later, the NBS was used to practise various 'blue-collar' construction jobs prior to the programme of assembling the first space stations.

The astronaut is submerged fully-suited, with weights attached to maintain a stable position in the water. Floating at a pre-determined depth, he can experience conditions very similar to those in space. Professional divers with scuba gear oversee the training. Several tanks are in use around the world.

A Skylab astronaut, (top), carries out a Neutral Buoyancy Simulation Test in preparation for Skylab EVA manoeuvres. These involved retrieving film and experimental equipment. Above: two Skylab astronauts are assisted by scuba divers during an NBS test. These exercises provided time and motion data as well as training for the crew in fully pressure-suited second by second conditions.

Training Simulators(2)

Ballistic Trajectory — zero-G

One of the oldest methods of providing brief periods of free-fall is by means of an aircraft flying a parabolic trajectory. This kind of simulation is expensive, requiring special flights in a modified aircraft, and the maximum period of weightlessness is little more than half a minute. Nevertheless, it is the most effective means of achieving a weightless condition short of actual orbital flight. The effect when the aircraft is put into an up-and-over parabola is similar to that in a rapidly descending elevator.

Horizontal Walking Device — reduced gravity

In preparation for planetary EVA (see p20-21), it is essential to be able to simulate actual conditions. The earliest system enabling an astronaut to experience lunar gravity, which is one-sixth that of the Earth, was developed at NASA's Langley Research Center. It consisted simply of a system of slings to support the astronaut horizontally. The angle of the slings to the perpendicular determined the G-effect, and a vertical wall was used as the 'floor'. This rather primitive system is no longer used.

Six Degrees of Freedom — reduced gravity

The Six Degrees of Operational Freedom Simulator uses a system of counter-balances (tailored to each astronaut's suited weight) to simulate lunar gravity or any low gravity condition. The astronaut is strapped in and must perform his training task whilst following the rotation of the simulator, unless the device has been fitted with a mobile tracking unit. Lunar conditions are reproduced with tolerable accuracy in terms of actual effort expended and force required, if not in terms of the wider experience, which is far more bizarre than anything the Moon has to offer.

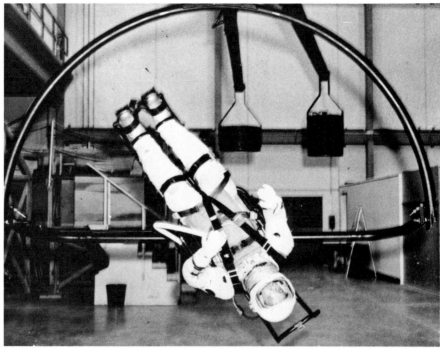

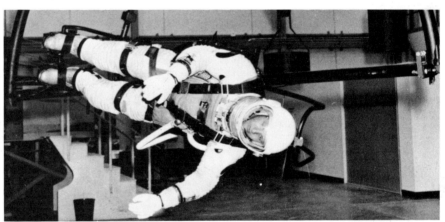

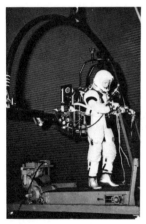

Above: the six degrees of operational freedom simulator is used for metabolic testing on a treadmill in a Life Systems Laboratory.
Right and top right: the same apparatus enables an astronaut to 'float' in different positions. The difficulty of using conventional tools in zero-G is illustrated bottom right.
Left: astronauts practise working fully suited in zero-G in the modified cabin of an aircraft in parabolic trajectory.
Top left: the rather basic horizontal walking device.

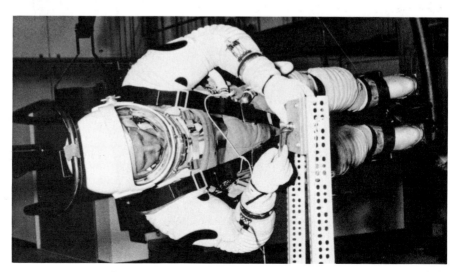

Training Simulators(3)

The Centrifuge — high-G

The centrifuge, the most disliked of all training devices, has a surprisingly long and august history. Erasmus Darwin, father of Charles, first proposed centrifugation in 1795 as a means of inducing sleep! This was followed by several practical attempts to cure various aspects of insanity.

Nowadays, among many other research functions, the centrifuge has an important role in creating the kind of G-forces that are experienced during lift-off and re-entry. Training can increase the capacity of subjects to withstand higher G-forces.

The Mastif — disorientation

A carnival ride with a purpose, the Multiple Axis Space Test Inertia Facility (Mastif) allows spinning on three axes of rotation, and is used to train astronauts to control wild tumbling. A spacecraft, or simply a suited astronaut on EVA, can easily start to tumble erratically, as neither gravity nor atmosphere are present to slow the motion. The immediate danger is motion sickness, but 'three way tumbling' can also be most disorienting, and without practice can be very difficult to correct.

The astronaut sits in the inner of three cages, one inside the other. Each cage is spun on an axis perpendicular to that of the other two.

Control systems in the Mastif, operated by a single joy-stick, allow the astronaut, strapped into a couch, to counter the tumbling effect by using spurts of nitrogen gas. Tumbling is normally operated at up to 30 rpm in each direction.

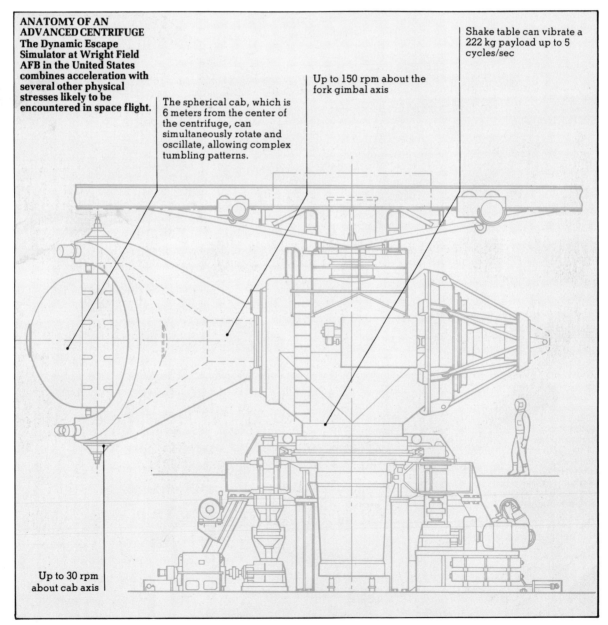

ANATOMY OF AN ADVANCED CENTRIFUGE
The Dynamic Escape Simulator at Wright Field AFB in the United States combines acceleration with several other physical stresses likely to be encountered in space flight.

The spherical cab, which is 6 meters from the center of the centrifuge, can simultaneously rotate and oscillate, allowing complex tumbling patterns.

Up to 150 rpm about the fork gimbal axis

Shake table can vibrate a 222 kg payload up to 5 cycles/sec

Up to 30 rpm about cab axis

Left: the famous US Navy Johnsville Centrifuge, built in 1970, was used throughout the early US Space Program, and in its time was the most powerful in existence. With a 17 meter radius it had a rate of change of 10G/sec and could reach 40G/sec. The 10 meter-diameter gimbal-mounted chamber was fully air-conditioned.
Below: The Mastif consists of three cages, one inside the other. The astronaut sits in the inner one, affected by the spin of all three.

Medical The body's needs

THE PHILOSOPHY OF SPACE environments has moved over the years towards the view that living conditions in space should be as similar as possible to those on Earth. Early hopes that short-cuts could be made for astronauts, and that they could exist in space without faithfully reproducing *every* condition, came to little. Experience has shown that over long periods even seemingly trivial things can be important. However, exact conditions are not so crucial on short journeys, and there is often cost-saving here.

Atmosphere

The most immediate human need in space is an atmosphere of acceptable composition and pressure. Both factors can vary, because what matters is the partial oxygen pressure (pO_2), that is the pressure of O_2 in the atmosphere, irrespective of other gases present. pO_2 at sea-level on Earth is 22,700 Pa (Pa = pascal, a unit of pressure) or 3.1 pounds per square inch. It is now generally accepted that 9000 Pa higher or lower than this marks the safe range. At less than about 13,400 Pa the lungs cannot function properly. At about 32,000 Pa, the distribution of micro-organisms in the blood can change drastically.

Pure oxygen is quite dangerous, not only because of fire risk, but also because it would result in decompression in spaces such as the sinuses. An inert gas is needed, and not surprisingly it turns out that nitrogen is the most satisfactory. Eighty per cent of the Earth's atmosphere is nitrogen, so many organisms need it for proper development. It is interesting to look at the differences in the atmospheres provided by the Russians and Americans in the early days. Spacecraft that can withstand full atmospheric pressure are heavier than those pressurized to less. The Russians started with an extremely powerful launch vehicle, the A-1, and were able to provide a normal atmosphere for their cosmonauts. The Americans opted for a much lower cabin pressure.

Partial CO_2 pressure must be less than 400 Pa, but not so low as to inhibit photosynthesis on board those vehicles using plant life in their life support systems (see p.34)

Humidity and temperature must be at comfortable levels, and these are reasonably flexible. Forty per cent humidity and 22°C are normal. Humidity is controlled by passing air through heat exchangers that cause condensation. Temperature is usually controlled by means of both active and passive systems; actual heating is derived from the spacecraft's power systems and/or solar panels, and is distributed through air-ducts. Cooling, which is rather more complicated, is provided by refrigeration plants, heat exchangers, and external radiators that dump excess heat into space. Close to the Sun, excess absorption of heat is prevented either by reflective shielding or by rolling the ship.

Gravity

The full importance of a normal Earth gravity can best be appreciated by looking at some of the bizarre effects that weightlessness causes. Calcium production for the bones cannot continue without the pressure that gravity supplies, and the calcium loss of 1-2 per cent a month

Right: a Dr Kerwin gives Skylab astronaut Conrad an oral examination.
Left: he monitors the performance of Paul J. Weitz, another Skylab astronaut, on the bicycle ergometer. This device was used to determine whether a man's effectiveness in doing mechanical work was affected by a prolonged stay in space – a vital experiment for the early Skylab astronauts.
Above: a typical medical kit. shows typical suffering

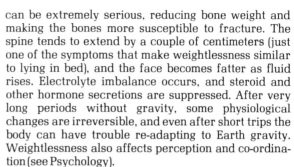

**EXPLOSIVE
DECOMPRESSION
THE ULTIMATE DANGER**

**Failure of the
pressurized environment
is every astronaut's fear.
Sudden exposure to a
vacuum would cause the
body to swell and
rupture as the internal
gases rapidly expanded,
and cause the blood to
boil. The famous case of
the Osiris II in 2037,
when the pilot of the
wrecked ship had to
make the short passage
to the rescue ship
without a pressure suit,
proves that survival for
a few seconds is
sometimes just possible.
In this case, the pilot
took the precaution of
hyperventilating to boost
the oxygen content of
his blood, and then
deflated his lungs
completely before
exposing himself to
space. In this way he a
avoided lung rupture
and prolonged the
oxygen supply. The blood
pressure sensor (above)
monitors changes of a
less dramatic but vital
nature.**

can be extremely serious, reducing bone weight and making the bones more susceptible to fracture. The spine tends to extend by a couple of centimeters (just one of the symptoms that make weightlessness similar to lying in bed), and the face becomes fatter as fluid rises. Electrolyte imbalance occurs, and steroid and other hormone secretions are suppressed. After very long periods without gravity, some physiological changes are irreversible, and even after short trips the body can have trouble re-adapting to Earth gravity. Weightlessness also affects perception and co-ordination (see Psychology).

It is now generally accepted that either simulated gravity or a strict regime of exercise is necessary. Gravity is simulated by rotating the space structure, and the inhabitants live on the inside surface of the circumference. However, the size of the structure is important because of another factor — the Coriolis force. On a rotating object, whether it be a space station or the Earth, the Coriolis force is a sideways force that acts perpendicularly to the velocity and axis of rotation. On the Earth it deflects the flow of wind, and the flow of water down plug-holes; on a space station it can cause motion sickness by creating cross-coupled accelerations in the ear's semicircular canals. To maintain 1G, a small space structure has to rotate faster than a large one, creating a stronger Coriolis force. Three rpm is about the comfortable maximum speed of rotation, so the larger a space station the better.

One further problem caused by weightlessness is in the ship's galley. Bits of food floating around the dining area after a meal may be untidy, but the much smaller aerosols that are also present are a real health hazard. Space pneumonia is a common illness in zero-G spacecraft, although aerosol extraction plants can reduce its incidence.

Acceleration
Although commercial shuttle lines subject their passengers to less than 4G at take-off and, for a very short period, the effects of *high* gravity have been important in the past. Adequate bodily support in the correct position (facing the direction of the force) is essential, and people suffering from certain respiratory illnesses and weak hearts are in some danger. Tolerance of high-G forces can be acquired, by the judicious use of a centrifuge, although this is not considered necessary for shuttle passengers.

Radiation
The problems of radiation are dealt with in detail under Hazards (p. 154). Spacecraft shells are designed to provide protection against normal background radiation, whilst a central radiation shelter (usually designed so that it can be dismantled when not needed) is used for the big solar flares, which can be in the order of 12,000 BeV. The Colonies and major space stations use compressed lunar rubble fixed in sheets around their surfaces. Experimentation continues towards the ideal method, the plasma shield, providing automatic electromagnetic shielding. Travellers will know that the more experimental propulsion systems have proved a hazard, but public lines are quite safe.

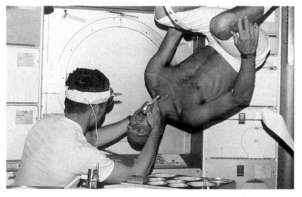

Psychology Possible disorders

THE VERY NATURE of long manned Space voyages, where typically a small group of individuals is confined in a highly ordered, isolated and predictable environment, provides a fertile breeding ground for psychological disorders. The experiences of the earliest astronauts — military-trained crew and goal-oriented scientists, on short flights — bear little resemblance to today's problems.

Self-medication is now fashionably encouraged by most practising consultants, and a short degree course in Space Psychology is highly recommended for those anticipating frequent off-Earth trips. The psychological complexities are such that there is no practicable method of equipping the astronaut for *all* space situations and psychotherapy is therefore always designed for specific missions.

Upset equilibrium
Weightlessness, unavoidable in many spacecraft designs, has a disturbing effect on the vestibular part of the inner ear, and this upsets the equilibrium. Although this is a physical symptom, it alters the ability of the astronaut to perceive normally, and can exaggerate psychological problems. Its effect can be permanent, as was the case with certain early American astronauts.

Many modern spacecraft have a low level of simulated gravity incorporated into their design.

Diminished stimulation
Weightlessness also reduces the tone and threshold of the central nervous system. This means that brain activity tends to diminish, possibly triggering apathy and fantasy. Here, again, simulated gravity is the ideal answer, but missions are usually designed so as to provide maximum mental stimulation.

The Solipsism Syndrome
Solipsism is a philosophical concept in which the only reality is that perceived by the observer — anything outside his direct experience is held not to exist. As a philosophical concept its nuisance value lies in the fact that it cannot be disproved, although commonsense says that it is inherently ridiculous. But, with the advent of space travel, this seemingly academic oddity has become a real psychological danger; the absence of stimuli in the regular, ordered environment of the spacecraft can easily persuade the astronaut that the real world begins and ends at the ship's hull. The longer the voyage, the worse the problem becomes.

To combat this, most missions incorporate a variety of in-flight activities for crew members, and these always involve observations of Space. Since the development of closed-cycle life support systems that use plant-life, the plants themselves have proved valuable in introducing growth and change into the crew's lives. Tending them is very popular.

Loss of Identity
Many of the symptoms and problems are interrelated, of course, and the increasingly long communications interval (messages from Earth to the Out-Planets can take hours to travel the distance) reinforces the solipsism syndrome by straining one more link with

reality. It also threatens the astronaut's sense of belonging, which can in turn stimulate loss of identity. Even at the quite neighbourly distance of the Moon, radio communications with the Earth contain nearly a second's lag, and this is sufficient to disjoint a conversation. Even with such a small interval, conversations lose their connectedness, and appear to the participants to be chopped into a number of separate one-way messages. At interplanetary distances many travellers report the unsettling feeling that the void is swallowing up their messages.

There is no answer to the communication gap, but the work programme helps maintain identity through providing a sense of purpose.

Fear
The existence of a hostile environment — Space — kept at bay by what can appear to be a fragile life support system, can gradually create anxiety. More crucial, however, is the inevitable mission structure, which intersperses very long periods of inactivity with short bursts of extreme activity, often under stress. This structure can produce fear. After perhaps months of regular and monotonous routine, when the ship is in its coasting phase, the crew must undertake planetfall, or at least planetary encounter. The contrast between these two mission phases is so extreme that 'target fear', as it is commonly known, is not infrequent as the spacecraft draws near its objective.

An accident or disruption can have the same effect; the suddenness does not allow the slow build-up characteristic of 'target fear', but the very nature of the crisis can cause shock and withdrawal.

Even though the periods of fear are well-defined and limited, drug treatment is out of the question, as the side-effects would affect alertness. One solution is to stress the goal-orientation of the mission. By encouraging the crew to focus on the encounter phase, it is hoped that their interest and involvement in their objective will overcome their fear of it. Additionally, support therapy can be provided by ground control, and there has been some recent experimentation with pre-programmed support therapy.

Sexual Frustration
The withdrawal of normal sexual outlets on long voyages (to date, there have been no male/female crews to the Out-Planets for fear of group stress and sexual jealousy) naturally invites historical comparison with forced single-sex communities, such as prisons. The importance of the problem depends largely on the sexual drives of the individual, but the key is to make sure that the inevitable sublimation is in harmless directions. Generally speaking, a full and varied work programme throughout the mission is an effective antidote. The drug Benpelidol apparently has no side effects of any importance, but many space travellers object on principle to a drug-reliant approach.

Claustrophobia
Pre-flight testing reveals claustrophobic tendencies, so that it is rarely encountered on board. However, severe experiences can induce it.

Food The fastest fast food

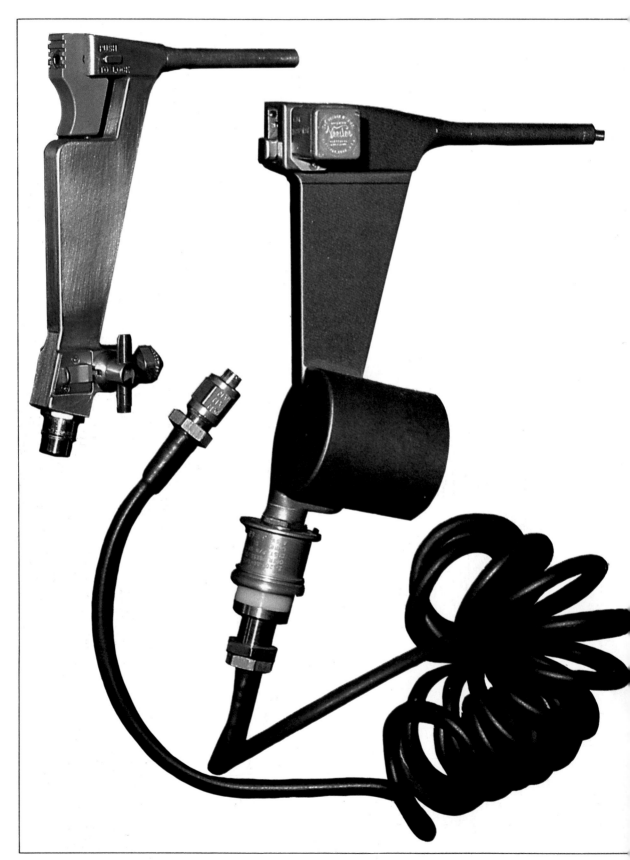

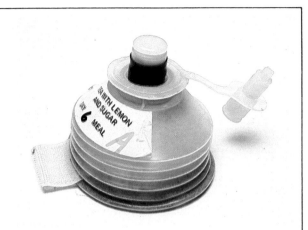

The standard drink pack makes the best of a tricky problem

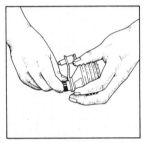

1. Break the seal on the nozzle and remove the top. At this stage the tea is a brown powder.

2. Put the water gun into the nozzle and insert the correct dosage of hot water (63°C); this will stretch the concertina pack.

3. Shake the pack to mix the drink thoroughly; the nylon valve prevents the water from leaking out.

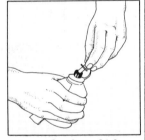

4. Insert the plug, which opens the valve for drinking.

5. Raise to mouth, and drink by either sucking or by squeezing the end of the concertina pack.

6. In between drinks, attach the Velcro tab to a matching tab on the tray.

ALTHOUGH IT HAS BEEN possible for many years to concentrate highly nutritious food into a small space, by such methods as freeze-drying and chemical synthesis, the main problem has always been that the food must be acceptable to astronauts. On short, specialized missions, astronauts can accept the limitations quite easily, knowing that an abnormal diet will only be temporary. On longer voyages, however, normal foods become very important, but unfortunately a normal Earth diet is weighty, and therefore costly to launch.

So, most of the technological work has been directed at packaging methods which will eventually deliver to the space traveller a relatively normal meal. On the first American Mercury flights, semi-solid 'baby food' was the staple, eaten through a tube system. Although convenient to eat, these purées were not liked, and this approach to food was abandoned long ago.

Rehydratable foods

Where weight is a serious problem, dehydrated foods are normally used. With their moisture gone, they are reduced to about one-third of their original weight. Freeze-dried and vacuum-packed, they can in an emergency be eaten dry, but under average shipboard conditions water (extracted and purified from body waste and fuel cell processes) is injected from a dosage 'gun' into standard nozzle valves on each

The water gun on the left (facing page) dispensed hot water, the one on the right dispensed cold water. The fuel cells created electricity by combining hydrogen and oxygen, and the water was produced as a by-product of this process. Some was chilled for drinking, the rest was heated for hot meals. Left: the Gemini water gun reconstitutes space food. Above: a Gemini astronaut drinks orange juice — probably a more enjoyable experience than eating by the same method.

THE BODY'S MEAGRE NEEDS
Centuries of complex eating patterns and the rise in status of eating from a means of staying alive to a social activity have both masked the body's actual needs. Most of our diet supplies only calories for body activity, and the essential components make up less than 10 per cent of the weight of our food. On average, we need only 35g/day of protein, 2-5g/day of polyunsaturated fat, 5-10g/day of salts, trace elements and vitamins, and a tiny amount of ordinary carbohydrate (see left). Now that closed-cycle life-support systems are entering into common use, where human waste products such as carbon dioxide, water and urea are converted back into oxygen and purified water, most space travellers have become accustomed to what would once have been considered a repulsive idea. It is now psychologically possible (the technical side of it is less of a problem) to introduce chemical synthesis of food on long flights. Formaldehyde can be converted into formose sugars, and fatty acids, glycerol and fat were synthesized as long ago as 1944 in Germany.

packet. If the food has simply been moistened rather than thoroughly wetted, it can be eaten with a spoon even in zero-gravity.

Thermostabilized moist foods
Where weight permits, thermostabilized foods are supplied, in flexible packs with a normal moisture content. These are every astronaut's favourite, closely resembling normal food. Spreads and fillings are available in small aluminium tubs with ring-pull tops, and on the shorter voyages these can be used with fresh bread packed under a slight pressure of nitrogen (which stops it drying too much and becoming crumbly).

Eating in zero-G — no room for bad table manners
After all the problems of preparing and packaging a space meal comes the difficulty of eating it. In spacecraft without artificial gravity, the astronaut must resort to various devices to stop the food floating away. Liquids tend to form globules, breaking into smaller droplets when touched and a cup is of no value at all (the drink will just float out!). Loose food has no reason to stay on the plate, and readily finds its way to the grilles of the air extraction system. Even sitting at the table poses a problem, and most dining room chairs in space are equipped with seat-belts.

The small container of bite-sized strawberry cereal cubes (above) formed part of the original Space Shuttle cold menu. Preparation and packing facilities were tested at the NASA kitchen, (right). These specially-equipped kitchens marked the end of sub-contracting and the start of in-house food preparation.

Waste management What nature intended?

URINATION AND DEFECATION in zero-gravity, or waste management as they are collectively and euphemistically known, are a somewhat delicate problem. Only a space traveller can appreciate the crucial help that gravity provides in Earthbound washrooms. In space, other devices are needed to direct the flow.

On board spacecraft, the urine and fecal systems are separate, each requiring its own solutions. Urine collection is relatively simple, involving only a liquid. Each astronaut has his own relief tube (varying according to the size of the individual); in the case of female astronauts, the attachment is rather more complicated and certainly less comfortable. As a substitute for gravity, either a vacuum or a fan are used at the far end of the urine collection dump — otherwise, of course, the urine would simply pool in the relief tube. In a vacuum system, a vacuum can be created by an adjustable valve opening outside the spacecraft. It is quite normal to dump urine directly out into space, where it instantly flashes into ice crystals. These reportedly form a beautiful halo around the ship, particularly when backlit by the Sun. When a fan is being used, or when the urine must be collected for medical examination, a hydrophilic filter is necessary. This is in the form of a bag halfway along the relief tube that passes air but not liquid.

Fecal collection

This is a long process not to be undertaken casually by beginners. The lavatory seat is specially contoured to make a tight fit and, in the absence of gravity, is not necessarily on the 'floor'. Two handles on either side enable the user to hold on tight. The next problem is a delicate one, with a most ingenious solution. Without gravity, there is no natural way for the bolus to separate from the anus. The answer has been to direct tiny jets of air from around the circumference of the actual lavatory seat towards the exit point, and these cause the separation. This method is made particularly easy to construct, because the partial vacuum inside the fecal collection assembly draws these air jets in automatically.

After this an airflow, generated by either vacuum or fan, just as in the case of urine, carries the feces down the system. It is normal to collect the feces for medical examination, and this is done in a bag measuring 30cm by about 15cm which, being hydrophilic, lets the airflow pass through it but retains the solids. The principle is analogous to a vacuum cleaner, to which the exploded view (p. 41) bears some resemblance. A vacuum oven is later used, when the fecal bags have been removed, to dry off all the liquid, as in the preparation of some dehydrated foods. Lavatory paper is supplied in sanitized bags, but its use is rather more difficult than most people imagine. The Skylab astronaut Conrad is first credited with the mirror that is normally installed on a double-jointed arm for more accurate wiping. This was naturally called the 'rear-view mirror'.

In some spacecraft where there is insufficient room for the normal washroom, a more primitive system is used, known as a contingency fecal bag. This is a one-

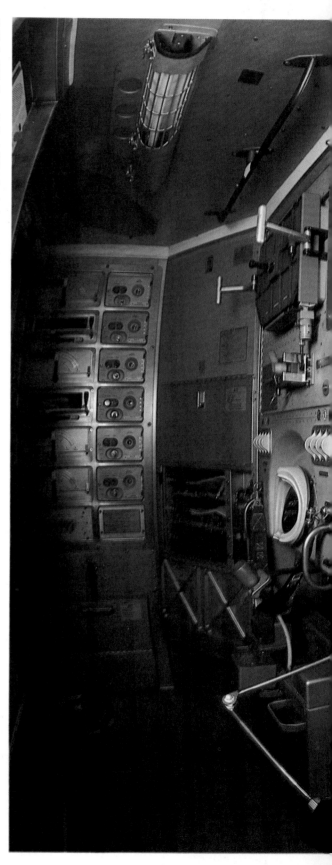

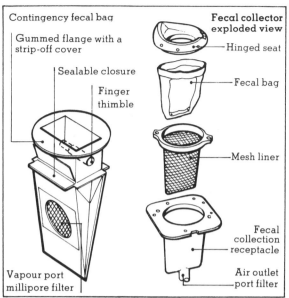

Contingency fecal bag

Gummed flange with a strip-off cover

Sealable closure

Finger thimble

Fecal collector exploded view

Hinged seat

Fecal bag

Mesh liner

Fecal collection receptacle

Vapour port millipore filter

Air outlet port filter

Left: the Skylab waste management compartment. The fecal-urine collector is mounted on the wall. The apparatus incorporated equipment for sampling and preserving certain body wastes. The return of samples to Earth was necessary to determine the biological effects of space-flight. The compartment also included utility closets. Crewmen could urinate from the standing or sitting position. Al Bean (right) adopts the standing one.

time-use plastic bag, which the user fits by sticking it on (it has an adhesive cover). Naturally, with no vacuum tube or air jets to ease the process, clearing the feces away from the body is a manual operation, for which a finger thimble has been thoughtfully let into the side of the bag near the top. After use, the top is pushed inside the bag, which has an adhesive sealable closure.

Few EVAs last for more than about an hour, so waste management inside the suit is something of a last resort. Urination by males can be accomplished with a semblance of dignity, but all the operations are a firm reminder of the animal in us.

Once again, urine collection is the simpler of the two systems. For men, a relief tube is fitted on by using a roll-on cuff. Whilst entirely natural for reasons of male pride, it is a real mistake to choose a relief tube that is too wide. This, fitted with a one-way flexible valve, connects to a urine collection bag that can take rather more than one liter. The equivalent fitting for women is quite uncomfortable and often difficult to make secure.

Use of the fecal collection system inside a suit is definitely a last-ditch measure. As the name 'fecal containment system' implies, it consists only of a pair of stretchable non-permeable diapers that do nothing more than contain the feces where they are.

Photography Adapting your camera

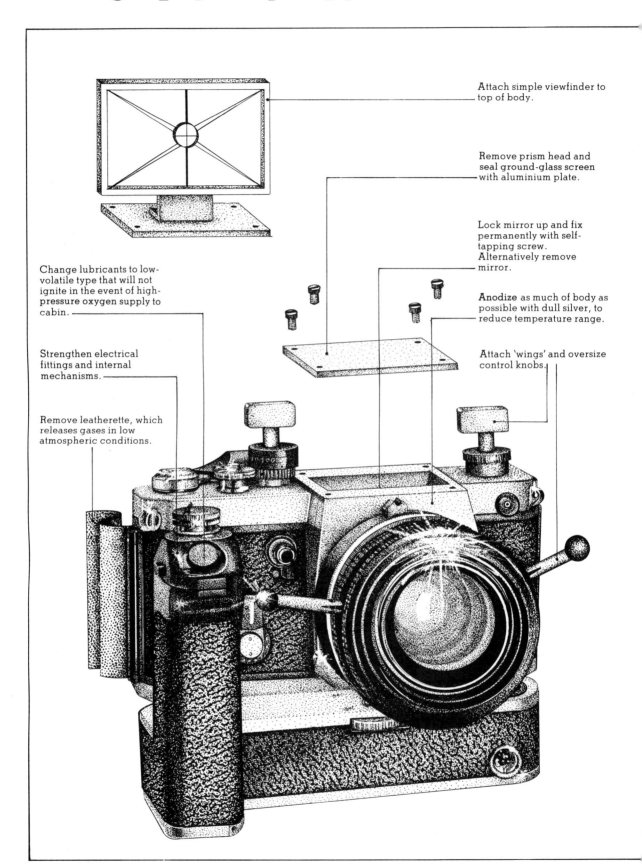

Attach simple viewfinder to top of body.

Remove prism head and seal ground-glass screen with aluminium plate.

Lock mirror up and fix permanently with self-tapping screw. Alternatively remove mirror.

Anodize as much of body as possible with dull silver, to reduce temperature range.

Attach 'wings' and oversize control knobs.

Change lubricants to low-volatile type that will not ignite in the event of high-**pressure oxygen supply to** cabin.

Strengthen electrical fittings and internal mechanisms.

Remove leatherette, which releases gases in low atmospheric conditions.

EXPOSURE — NOMINAL SETTINGS
Nominal settings for 64 ASA film on lunar surface and in Earth Space

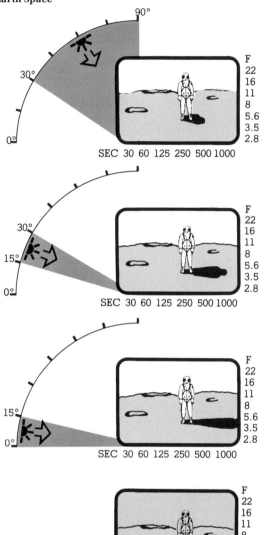

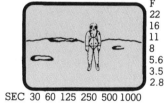

64 ASA film	Mercury	1/250 sec	f22
	Venus	1/250 sec	f16*
	Mars	1/250 sec	f8*
	Jupiter	1/125 sec	f3.5
	Saturn	1/60 sec	f2.8
400 ASA film	Uranus	1/125 sec	f3.5
400 ASA film rated 800 ASA	Neptune	1/60 sec	f3.5

*Use a meter on the surface. Under average climatic conditions, surface exposure on Venus should be about 1/30 sec at f2.8, and on Mars, without dust-storms, 1/125 sec at f5.6.

THE TRAVELLER MAY FEEL a little daunted at the thought of interplanetary photography. There is no need. With some thought and preparation, anyone who can turn out satisfactory holiday snapshots on Earth is assured of some minor pictorial masterpieces in space. The traveller will also find the camera an invaluable tool for making accurate scientific and engineering records.

Choosing the right equipment
It is a common fallacy that special cameras are needed in space. For years, the United States programme used conventional still cameras, such as Hasselblad, Minolta and Nikon, modified to NASA requirements. With a small soldering iron and watchmaker's tools, the camera owner can do most of the necessary modifications himself, particularly if the camera has a minimum of electronics. Every precaution must be taken to avoid glare from bright parts — this can cause permanent retinal damage. This is one reason for removing the prism head and mirror; another is the impracticality of using the normal SLR viewfinder with a space helmet. Attaching 'wings' to the controls will make handling easier when wearing a spacesuit. A matt silver anodized finish will protect the camera from overheating in sunlight and from heat-loss in shadow.

Selecting the right film
Professional astronauts use colour reversal film almost exclusively. Reversal film (transparencies) has better resolution than negative film, and removes the problem of guessing the colour of unfamiliar planets when making prints, which can in any case be made from transparencies. 64 ASA film, such as Ektachrome 64, is suitable for general photography from Mercury Space as far as Saturn Space. Beyond Saturn, natural light is insufficient for normal film, and a high-speed film, such as Ektachrome 400, is recommended. This can be rated to 800 ASA quite satisfactorily.

Getting the right exposure
What can at first seem to be an enormous stumbling-block is easily reduced to manageable proportions by the use of 'nominal settings', used extensively on the Apollo lunar missions. Although a little crude as a

The automatic spotmeter (left) was used on Apollo 11 in the first attempt to make accurate exposures in the then unfamiliar lighting conditions of space. The ASA film speed and camera shutter setting are first dialled in. The subject is observed through a telescopic eyepiece and a superimposed circle shows the small area which is being measured. The meter is switched on by pulling a trigger and the 'f' number for correct exposure lights up as a digital display. **Overleaf: You need not photograph Saturn in colour to capture its splendour.**

means of determining exposure, it has been found to work well in practice. In deep space, or on the surface of an airless moon, the contrast range between sunlight and shadow is very high, and exposure depends on the angle of the Sun. Exposure is also affected by the distance from the Sun. Travelling to the outer planets, each time the distance from the Sun is doubled, two extra stops are needed.

When really accurate exposures are needed, a spotmeter is recommended, with an acceptance angle of no more than 1°, and with a wide range of ASA speeds (the best cover 3 to 25,000).

One special problem that the traveller may encounter is accompanying ship debris, the product of engine burns and waste disposal, which has the annoying habit of forming a cloud of particles outside the spacecraft windows. The only answer — using thrust to separate the trajectories of ship and debris — expends valuable fuel.

Processing the film
Processing at any commercial laboratory can be normal, except when the film has been over-rated to compensate for low-light conditions. If the traveller suspects that the film has been exposed to excessive radiation, this should be reported to the laboratory, which will reduce development times according to pre-determined tables.

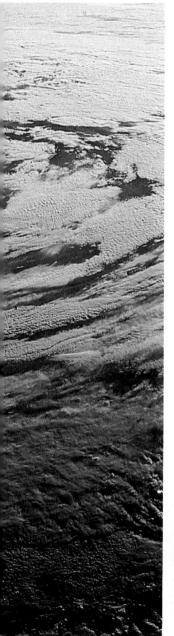

Left: Typhoon Bernice
photographed from Apollo
11. The sun angle was very
low in the sky.
Above: Atmospheric
layering, taken from Skylab.
Right: clouds cast shadows
on a lower cloud deck. The
sunlit horizon is top right,
the night/day divide top left.
Below: the Grand Canyon,
Painted Desert and Lake
Powell.

PHOTOGRAPHIC MEMORABILIA

The earliest pieces of
camera equipment used in
space exploration have
now become collectors'
items. All of these items
were used in the Apollo
Moon programme and are
still on display in
Washington.

Compact 16mm data-
acquisition camera (right)
clamped in one of the
windows of the Apollo 12
lunar module, used to
record the moonwalk.

A B/W TV camera with additional
lenses used on Apollo 7 in 1968.
This transmitted the first live TV from
space in October 1968.

Space Law Regulations without teeth

THE GROUNDWORK for the increasingly important laws governing space activities was laid down during the 1960s under the auspices of the Outer Space Committee of the United Nations. At the time, the United Nations was the natural organization to undertake this task, but the legacy of the early Treaties and Resolutions was for many, decades disastrous, reflecting the United Nations' traditional inadequacy.

Nevertheless, the lasting benefit of the UN's early involvement was its high-principled approach, assuming the freedom of outer space and an atmosphere of international cooperation and mutual benefit. Needless to say, whilst this approach was acceptable at a period when only research and exploration was being carried out, it fell rapidly by the wayside as soon as spacefaring nations began to develop resources for their own benefit. It was, in fact, such action that ended the United Nations' influence in space. This was the appropriation by Russia in 2004 of 500 sq km near Clavius Base, which demonstrated that the only means of enforcing the agreements, through international negotiations, was singularly inadequate. (The 1972 Convention made only the following provision for claims: 'A claim for compensation for damage shall be presented to a launching State through diplomatic channels (or) it may also present a claim through the Secretary-General of the United Nations.) It was as a direct result of this incident that the International Space Authority (the ISA) was founded, and this became the arbiter of legal disputes.

Space — a new legal setting

Although international lawyers had already had some limited experience with unexplored and hostile environments (Antarctica and undersea areas), space posed some entirely new problems. All bodies in space are entirely separate from the Earth and are thus virgin territory. Therefore, what has tended to happen is that the first nation to land, settle and exploit a particular site has a *de facto* claim to it, to the continuing frustration of the non-spacefaring nations of the world. The 1976 session of the UN committee showed the typical conflict of views — Chile talked of the 'common heritage of mankind' and claimed that 'such resources should not be the property only of those who are able to explore and exploit them', whilst Russia could 'not agree with the proposal of the concept of "the common heritage of mankind"', principally because it felt 'that such a proposal is premature in the absence of the necessary objective foundations and factual material for it'. This posturing of high-minded indignation versus legalistic stonewalling has been the monotonous theme of space legislation for more than a century. Consequently, most legal principles remain as vague as they were under the United Nations.

Fortunately, one thing that *has* been agreed is the legal definition of space. The 1968 ILA Conference at Buenos Aires established the principle that Outer Space starts at the altitude at which a spacecraft or satellite can orbit the Earth, and this was later defined as 100 km for the purposes of law. Unfortunately, it was also claimed, in the Bogota Declaration of 1976, that the special properties of the Geosynchronous Earth Orbit

KEY LEGISLATION AND EVENTS
Note: dates refer to the year the agreement came into force.

1958 UN General Assembly first considered space exploitation.
1958 UN Outer Space Committee created.
1961 UN General Assembly defines functions of Outer Space Committee.
1963 *Declaration of Legal Principles Governing Activities of States in Exploration and Use of Outer Space* (UN General Assembly): sovereignty, applications of international law, responsibility and liability, status of astronauts as envoys of mankind.
1966 International Law Association (ILA) Helsinki Conference: demarcation.
1967 *Treaty on Principles Governing the Activities of States in the Exploration and Uses of Outer Space, including the Moon and Other Celestial Bodies* (UN General Assembly): elaboration of 1963 Declaration, weapons ban, military restrictions, freedom of access to space vehicles.
1968 *Agreement on the Rescue of Astronauts, the Return of Astronauts and the Return of Objects Launched into Outer Space* (UN General Assembly): accident procedures.
1968 ILA Buenos Aires Conference: demarcation.
1971 Draft Moon Treaty (UN General Assembly).

50TH ANNUAL CONVENTION

(GEO) make it a scarce natural resource. Here, again, the lines were drawn between those nations *with* space capabilities and those without, but the resolution of the question in 2006 was the ISA's first achievement. The GEO is regarded as a special case and its segments are allocated by the ISA.

One important change of direction took place in 2019, when it was realized that nuclear pulse rockets were feasible. Article IV of the 1967 Treaty (2222/XXI), which specifically forbade the testing and use of nuclear weapons in space, was amended to allow the use of nuclear devices *'for peaceful purposes'*. This also paved the way for nuclear terraforming experiments.

Only nations take the responsibility

Some unusual anomalies have arisen due to the abiding principle, established in the 1963 United Nations Declaration (principles 5 and 8), that States bear responsibility for national activities in outer space and, most importantly, that the State from whose territory or facility an object is launched is liable for any damage it may cause. Articles VI and VII of the 1967 Treaty confirmed this. The anomalies started to occur when countries began to lease their launching facilities to other nations, and became worse when individuals and private organizations began to use these facilities. As early as 1967, the West German company OTRAG leased from Zaire's President Mobutu a section of Shaba province, from which it undertook commercial launchings for anyone with the price of the ticket. By granting sovereignty over this territory to OTRAG, Mobutu cleverly absolved himself from liability.

More significantly, in 2047 a Belt miner returning to Earth with a small asteroid collided with Uganda's reconditioned Soyuz. The defendant claimed that, as his craft had been launched from the French space centre at Kourou, he was not liable. The International Court of Space found for the defendant, but the Treaty of 2055 amended the offending Article.

International law is no law

The most unsatisfactory aspect of the whole legal question in space is that the effectiveness of international legislation depends entirely on the good will of nations. Not all nations are signatory to all treaties, some elements of international space law are plainly at odds with the national law of some countries, and in the final analysis a nation can simply ignore the findings of the International Court of Space.

Basically, international law, on Earth as well as in space, is a conflict of law, the confrontation of two nations, each with its own set of internal laws. Legislation must be by treaty, and legal disputes tend to follow diplomatic channels in the first instance. The setting up of the International Court of Space by the ISA was an attempt to regulate disputes, but its only means of enforcing its judgements is to present its recommendations to the ISA. Essentially, the only punishment is sanction, (such as was applied to Rhodesia after UDI). This is only effective if a sufficient number of nations agree to undertake it. Even criminal cases against individuals must in the end be referred to national courts.

1972 *Convention on International Liability for Damage caused by Space Objects* (UN General Assembly).
1974 ILA New Delhi Conference: telecommunications by satellite.
1976 *Convention on the Registration of Objects Launched into Outer Space* (UN General Assembly).
1976 *Bogota Declaration*: status of GEO.
1991 *Convention on Space Communications* (UN General Assembly): allocation of frequencies.
1995 *Moon Treaty* (UN General Assembly): international cooperation, exploitation, research segments.
2006 Establishment of ISA, with legal jurisdiction.
2006 *Treaty on Demarcation of Outer Space* (ISA).
2010 International Court of Space established at Geneva (ISA).
2019 *Treaty on General Activities in Outer Space and on all Planetary Bodies in the Solar System* (ISA): sovereignty, rules of astro-navigation, allocation of orbits, registration and control of space vehicles, nuclear testing.
2031 *Belt Charter* (ISA): rights of individuals in Asteroid Belt.
2055 *Treaty on Settlement and Colonization of Outer Space and All Extra-Terrestrial Localities within the Solar System (ISA)*: sovereignty, liability.

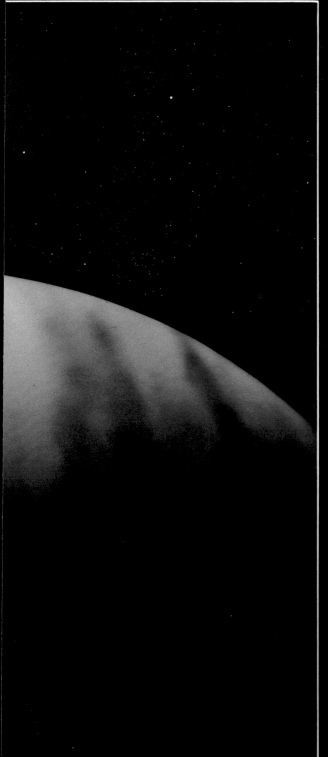

Hardware and spacecraft

IN A CENTURY AND A HALF of space travel, the progress of hardware has not been smooth. With the complete success of the Apollo programme, rocket engineers began to anticipate a steady growth of future exploration, with systems evolving to meet the next steps on the path to the planets, until finally Man embraced the entire Solar System. During the late 1960s NASA even prepared a blueprint for this future.

However, the commitment to a lunar landing having been fulfilled, the heavy expenditure began to be questioned — the more so in a new climate of economic recession and energy crises.

America pulled in its horns, and the waste of such expendable behemoths as the Saturn V was recognized. The second phase of exploration in the 1980s saw the beginnings of standardized re-usable vehicles and the financial justification of space travel — in short, *economic* hardware.

In the closing years of the 20th century, of course, men became more and more convinced of the immense possibilities of space, and space exploration became its own justification. Cycles of expansion and stagnation have controlled the development of new propulsion systems throughout the 21st century, and now once more there is bitter financial conflict over the enormous expense of crossing man's next threshold — interstellar space. It might be as well to remind those who argue against 'unjustifiable expense' that what seems now to be a difficult and unrealistic leap is in fact no greater than the settlement of the Solar System seemed after Apollo.

In geosynchronous orbit over Tycho Base, a lunar ferry unhooks its cargo of components for the Moon colony.

Pioneers The first rockets

THE PREHISTORY OF SPACE FLIGHT can, with some stretching of the imagination, be made to include the Chinese military fireworks of the first millenium AD, and the work of Congreve and Hale. Three landmarks, however, have practical significance in relation to modern spacecraft: the work of three men, Tsiolkovsky, Goddard and von Braun.

Without a formal education, and working entirely theoretically in Tsarist Russia, Konstantin Tsiolkovsky devoted his life to a visionary future of Man's colonization of space. Culminating in a paper entitled 'Exploration of Outer Space with Reactive Devices', Tsiolkovsky established basic principles of rocketry and space exploration that were far in advance of the technology of his day.

Unaware of Tsiolkovsky's theoretical work, the young American physicist Robert N. Goddard was looking for the practical solution to space flight. He

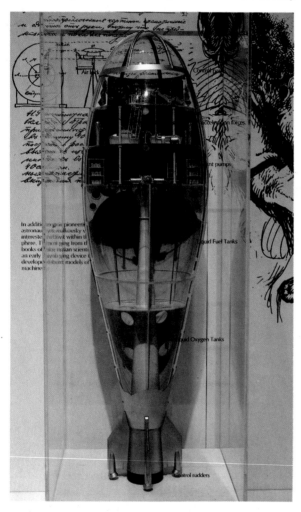

Tsiolkovsky was born near Moscow in 1857, and was totally deaf from the age of ten. He recognized that to travel in space men would need a sealed capsule supplied with the necessities of life.

He also realized that a reaction engine (see above) — the rocket — would be necessary to make the capsule travel through space. He was far in advance of his time in concluding that powder-propelled rockets could never achieve the speed necessary to reach space.

Goddard has been called the 'father of American rocketry' for being the first to experiment with liquid fuels, and for devising a single pump for both propellants (see rocket above right). Using gasoline and liquid oxygen he made the first successful launch in 1926 (see rocket above left).

He went on to introduce stabilizing, gyroscopically controlled vanes, and sophisticated techniques, such as making use of the rocket fuels to cool down the combustion chamber.

realized that a liquid oxygen/liquid hydrogen propellant offered the most promise, and, in the face of a general apathy and occasional derision, finally succeeded on 16 March 1926 in launching the world's first liquid propellant rocket.

The man who can truly be considered to have welded early rocketry to the actual exploration of space is Wernher von Braun, and the vehicle that linked the two eras was the German V-2 rocket. The Berlin rocket enthusiasts of the 1920s found backing from the German Army, and under the Nazis the well-developed state of the art was rigorously applied to weaponry. At the end of July 1944 the V-2, the world's first long-range strategic missile, was launched. After World War II, V-2 technology merged into the American rocket programme, and von Braun himself became Director of NASA's Marshall Space Flight Center for the programme that eventually put two men on the Moon.

Wernher von Braun's early work was on behalf of Hitler's government. At the end of WW 11 he engineered capture by the Americans.

The first US-modified V-2 was launched on April 16 1946. The V-2 (in stages of launch above) was 14 meters high and weighed about 12 tonnes, 2/3 of which was fuel and oxidizer. It burned alcohol and liquid oxygen For high-altitude stabilization, gyroscopes controlled internal vanes which could redirect the rocket blast.

HOW THE CONVENTIONAL ROCKET MOTOR WORKS

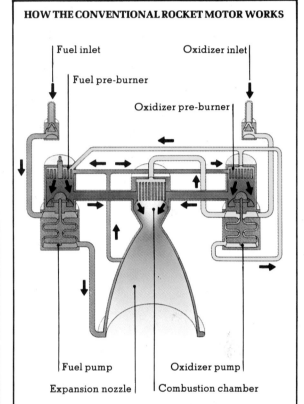

Fuel inlet

Oxidizer inlet

Fuel pre-burner

Oxidizer pre-burner

Fuel pump

Oxidizer pump

Expansion nozzle

Combustion chamber

The basic principle of the rocket is Newton's Third Law of Motion: 'to every action there is an equal and opposite reaction'. By expelling mass from one end at high speed, the rocket moves in the opposite direction. The velocity with which the rocket reacts depends on the quantity of 'reaction mass', as it is called, ejected and on the speed at which it leaves the nozzle. In the case of the basic chemical rocket, the propellant provides the reaction mass, and the mechanism by which this is expelled is combustion. The fuel (such as hydrogen or kerosene) and oxidant (such as liquid oxygen, or LOX) that make up the propellant burn together in the combustion chamber and release a high-speed stream of exhaust gases.

The speed at which these gases leave the rocket is called the *exhaust velocity*. The actual accelerating force that the motor develops is called *thrust*: this is the combination of the exhaust velocity and the quantity of reaction mass expelled in a given amount of time. Normally measured in tonnes (1 tonne = 1000 kg) the thrust must at least equal the original weight of the rocket if it is to leave the ground. Of course, in the absence of gravity, a low thrust will simply mean that it takes longer to reach the final velocity.

Early systems Pre-Apollo

THE FIRST PHASE of space exploration lasted twenty years and was wastefully conceived around expendable hardware. In many ways this was inevitable as the early rockets were adapted military missiles. These had been designed with the one objective of delivering a single payload to a target.

The other feature of this period was the commitment to a highly political goal. In the case of the United States, the goal that dominated their space programme was to put a man on the Moon. After Russia's achievement of being first to put a man into orbit, a manned lunar landing was felt to be the best way of demonstrating American dominance to the world.

America's first manned spaceflight was the launch of John Glenn in a Mercury capsule on 20 February 1962. The launch vehicle was an Atlas D from the US Air Force. In a way, the Mercury flights were the culmination of the very earliest American space efforts, set in motion before Kennedy made his famous commitment to a lunar landing.

After Mercury NASA realized that, in order to rehearse all the different stages of the final Moon landing, it would need to put a larger, two-man, capsule into orbit and be able to sustain it for up to a fortnight. The Gemini flights were the result in 1965 and 1966. The chosen launch vehicle was another military ICBM, the Titan II, with nearly four times the payload capacity of the Atlas.

The Americans hoped to put a man on the Moon using a lunar-orbit rendezvous and docking of lunar and command modules. In the Gemini mission (above) these manoeuvres were tried out successfully. It was also felt that on a long flight repairs outside the spacecraft could be necessary. During his 21-minute walk Ed White (left), showed that useful work could be done outside the craft. He enjoyed the walk so much that he had to be begged to return inside by his companion McDivitt.

The Mercury-Redstone rocket (below) was a modified Redstone missile topped with a one-man cone-shaped capsule. Its first successful launch was on 19 December 1960, through a single 35 tonne thrust rocket motor.

The Mercury-Atlas (below) was first launched successfully on 21 February 1961. It was planned that the Atlas would put a man in orbit after an initial up and down flight. Its two rocketdyne boosters gave a combined thrust of 140 tonnes.

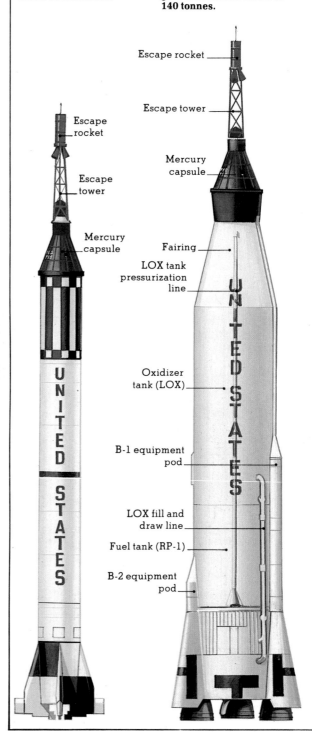

Escape rocket

Escape tower

Mercury capsule

Escape rocket

Fairing

Escape tower

LOX tank pressurization line

Mercury capsule

Oxidizer tank (LOX)

B-1 equipment pod

LOX fill and draw line

Fuel tank (RP-1)

B-2 equipment pod

Weighing 1360 kg and measuring 3 meters from heat shield to nose and two meters at its widest point, the Mercury capsule was about one third the size of Vostok.

The two-man Gemini capsule (below) was the first space ship that could be manoeuvred to change its orbital plane. Its centre of gravity was placed so that in certain positions it developed lift.

The combined Gemini spacecraft and Titan II rocket (right) gave 195,450 kg thrust at lift-off through its first-stage engine, and another 45,450 kg from its second stage to achieve orbit. *Overleaf:* **NASA's Marshall Space Flight Center at Huntsville Alabama. In the foreground is the original Redstone test stand, marking the beginning of America's manned space flights programme. In the background is the Space Shuttle Assembly building, marking the start of the era of cost-effective space flight.**

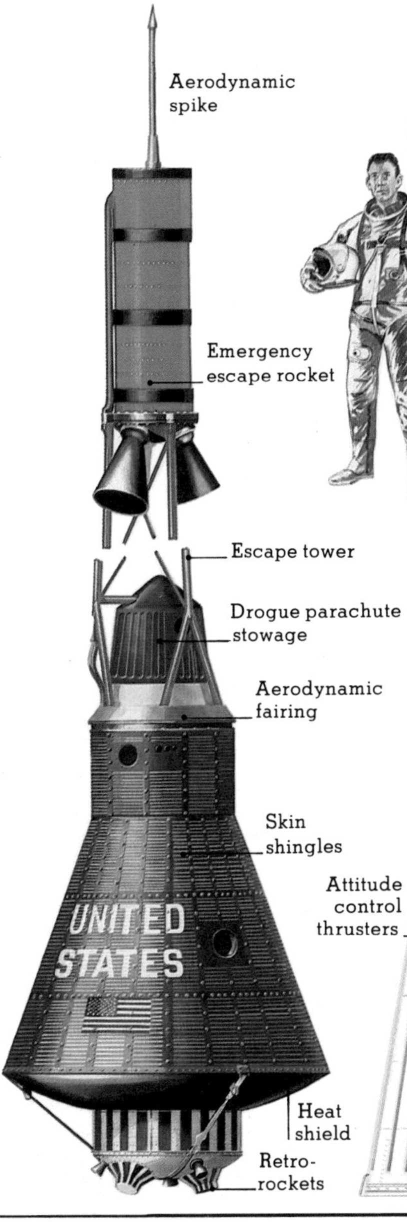

Aerodynamic spike

Emergency escape rocket

Escape tower

Drogue parachute stowage

Aerodynamic fairing

Skin shingles

Attitude control thrusters

UNITED STATES

Heat shield

Retro-rockets

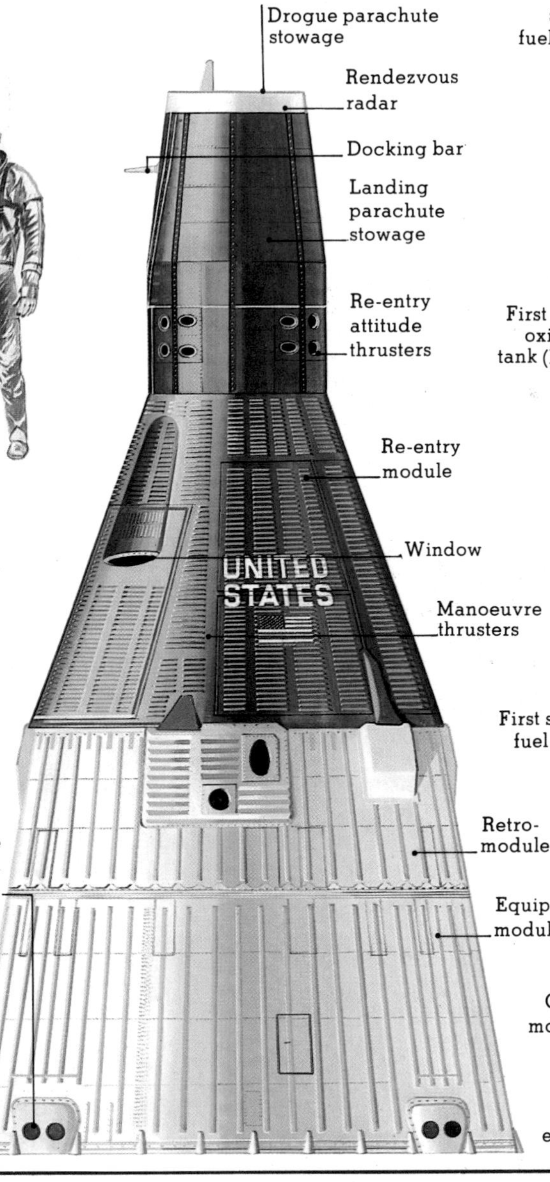

Drogue parachute stowage

Rendezvous radar

Docking bar

Landing parachute stowage

Re-entry attitude thrusters

Re-entry module

Window

Manoeuvre thrusters

Retro-module

Equipment module

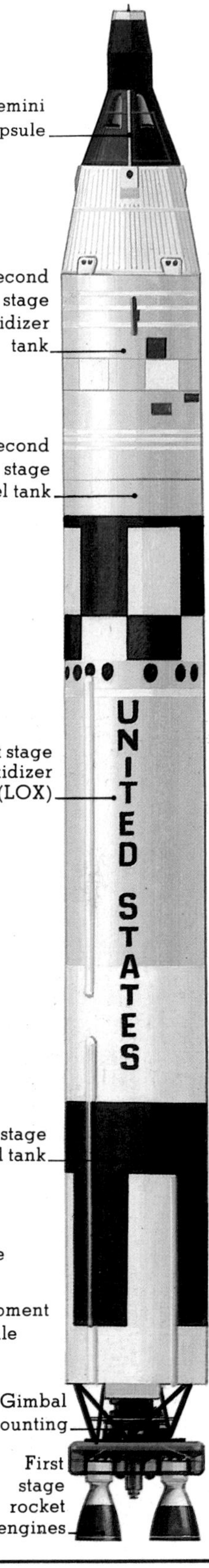

UNITED STATES

Gemini capsule

Second stage oxidizer tank

Second stage fuel tank

First stage oxidizer tank (LOX)

UNITED STATES

First stage fuel tank

Gimbal mounting

First stage rocket engines

Early systems Apollo

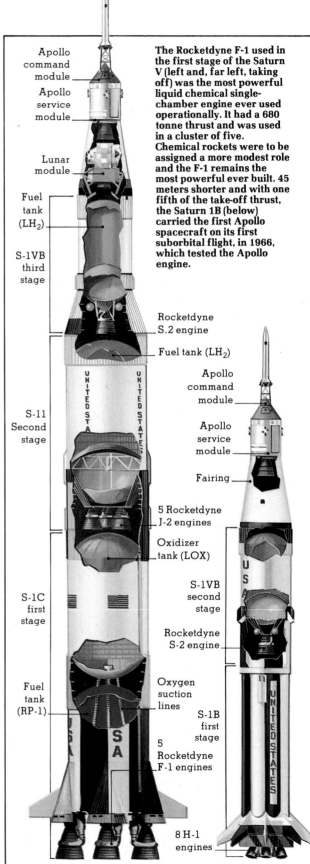

Apollo command module

Apollo service module

Lunar module

Fuel tank (LH₂)

S-1VB third stage

Rocketdyne S.2 engine

Fuel tank (LH₂)

S-11 Second stage

5 Rocketdyne J-2 engines

Oxidizer tank (LOX)

S-1C first stage

Apollo command module

Apollo service module

Fairing

S-1VB second stage

Rocketdyne S-2 engine

Fuel tank (RP-1)

Oxygen suction lines

5 Rocketdyne F-1 engines

S-1B first stage

8 H-1 engines

The Rocketdyne F-1 used in the first stage of the Saturn V (left and, far left, taking off) was the most powerful liquid chemical single-chamber engine ever used operationally. It had a 680 tonne thrust and was used in a cluster of five. Chemical rockets were to be assigned a more modest role and the F-1 remains the most powerful ever built. 45 meters shorter and with one fifth of the take-off thrust, the Saturn 1B (below) carried the first Apollo spacecraft on its first suborbital flight, in 1966, which tested the Apollo engine.

AFTER BITTER ARGUMENTS among the different parties involved in the space programme, it was finally decided at the end of 1962 that the Moon landing would be achieved by means of a Lunar Orbit Rendezvous (LOR). This determined the final configuration of the spacecraft: one single rocket had to deliver a double payload to lunar orbit — a relatively small Apollo module and a separate module that would land on and later ascend from the Moon's surface. This called (for the time) for a gigantic launch vehicle, one that could deliver a 47-tonne (47,000 kg) payload not just into Earth orbit but to the Moon.

The launch vehicle that was designed to meet this specification was the famous Saturn V, until the Russian G-1 the most powerful rocket ever built. In its

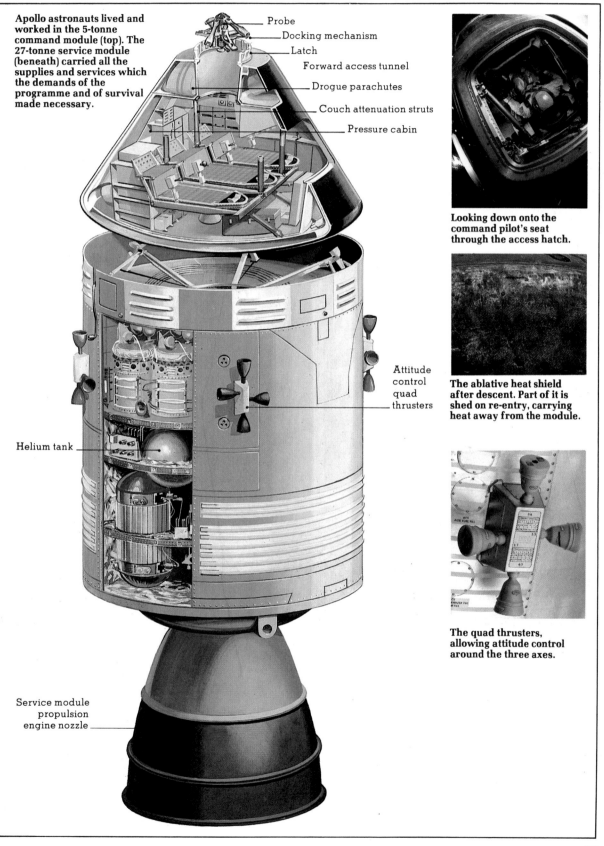

Apollo astronauts lived and worked in the 5-tonne command module (top). The 27-tonne service module (beneath) carried all the supplies and services which the demands of the programme and of survival made necessary.

Probe

Docking mechanism

Latch

Forward access tunnel

Drogue parachutes

Couch attenuation struts

Pressure cabin

Attitude control quad thrusters

Helium tank

Service module propulsion engine nozzle

Looking down onto the command pilot's seat through the access hatch.

The ablative heat shield after descent. Part of it is shed on re-entry, carrying heat away from the module.

The quad thrusters, allowing attitude control around the three axes.

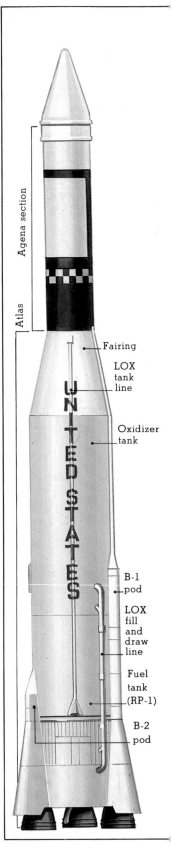

seven year life, the Saturn V accomplished six Moon landings and delivered one Skylab space station into Earth orbit.

Of the payload, only the Apollo Command and Service Modules are shown and the lunar module is dealt with on (p.100). After the last Apollo Moon landing in December 1972, Apollo craft were used in Earth orbit as a ferry for Skylab and in the experimental link-up with a Russian Soyuz.

Naturally, all hardware components from the Apollo programme are obsolete and are highly regarded as collectors' items. Amateurs are warned against using components of this class operationally; they are incompatible with modern systems and are unlikely to be in spaceworthy condition.

Right: the Atlas-Agena was 27 meters tall and weighed 125 tonnes at launch. On its first successful flight in May 1960 it carried a payload of 2.27 tonnes into orbit. The total thrust from the Atlas booster engines was 166 tonnes.

Top: David Scott steps into space from Apollo 9 — the Apollo flight which tested the lunar module in Earth orbit, and the independent life support system (PLSS) on a spacewalk.

Above: Apollo 17 shows its open SIM bay.

Agena section

Atlas

Fairing

LOX tank line

Oxidizer tank

UNITED STATES

B-1 pod

LOX fill and draw line

Fuel tank (RP-1)

B-2 pod

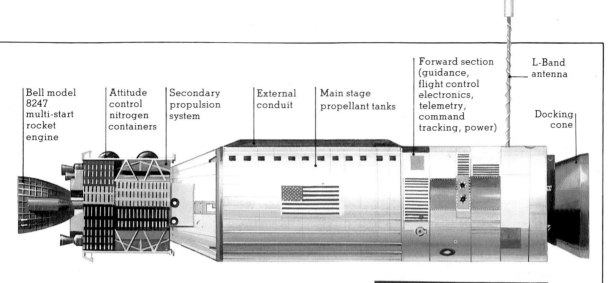

Bell model
8247
multi-start
rocket
engine

Attitude
control
nitrogen
containers

Secondary
propulsion
system

External
conduit

Main stage
propellant tanks

Forward section
(guidance,
flight control
electronics,
telemetry,
command
tracking, power)

L-Band
antenna

Docking
cone

The Agena D was modified to serve as a target vehicle for the manned Gemini spacecraft. An Atlas sent the Agena D into orbit to await the two-man crew. The crew moved the nose of their spacecraft into a special docking collar (below) on the Agena in a rehearsal of docking operations anticipated for the Apollo Moon landing. Six Gemini Agena target vehicles were sent up, with two failures.

The 'angry alligator' which prevented Gemini 9 from docking with its target vehicle was in fact a nose fairing covering the docking collar.

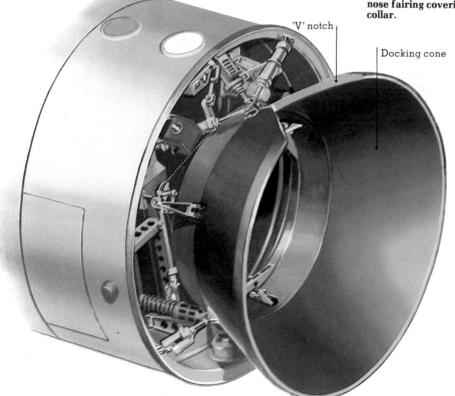

'V' notch

Docking cone

Early systems Vostok and Voskhod

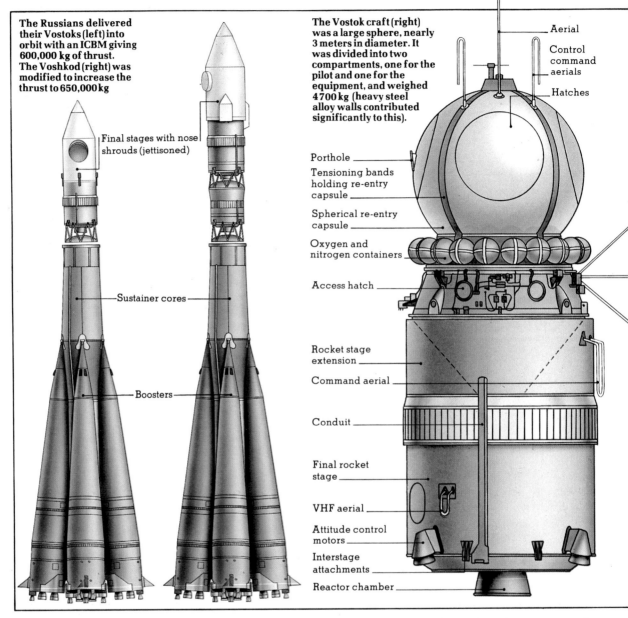

The Russians delivered their Vostoks (left) into orbit with an ICBM giving 600,000 kg of thrust. The Voshkod (right) was modified to increase the thrust to 650,000 kg

Final stages with nose shrouds (jettisoned)

Sustainer cores

Boosters

The Vostok craft (right) was a large sphere, nearly 3 meters in diameter. It was divided into two compartments, one for the pilot and one for the equipment, and weighed 4700 kg (heavy steel alloy walls contributed significantly to this).

Aerial

Control command aerials

Hatches

Porthole

Tensioning bands holding re-entry capsule

Spherical re-entry capsule

Oxygen and nitrogen containers

Access hatch

Rocket stage extension

Command aerial

Conduit

Final rocket stage

VHF aerial

Attitude control motors

Interstage attachments

Reactor chamber

RUSSIA'S BEGINNINGS in space were mostly shrouded in characteristic secrecy and mission intentions were not always made clear. The mission failure rate was undoubtedly higher than the Americans', but there were also some notable successes.

To an even greater extent than the United States, Russia based its space programme on military hardware. The parent launch vehicle for many years was the highly successful SS-6 ICBM, at one time the most powerful rocket in the world.

On 12 April 1961 a Vostok craft on an A-1 launcher carried Yuri Gagarin into orbit — the first man in space. The Vostok programme continued until 1965, when it was succeeded by two Voskhod flights. Voskhod and its launcher were, in essence, uprated versions of the earlier flights, enabling three astronauts to be carried into orbit.

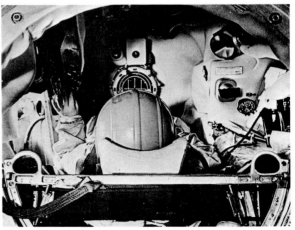

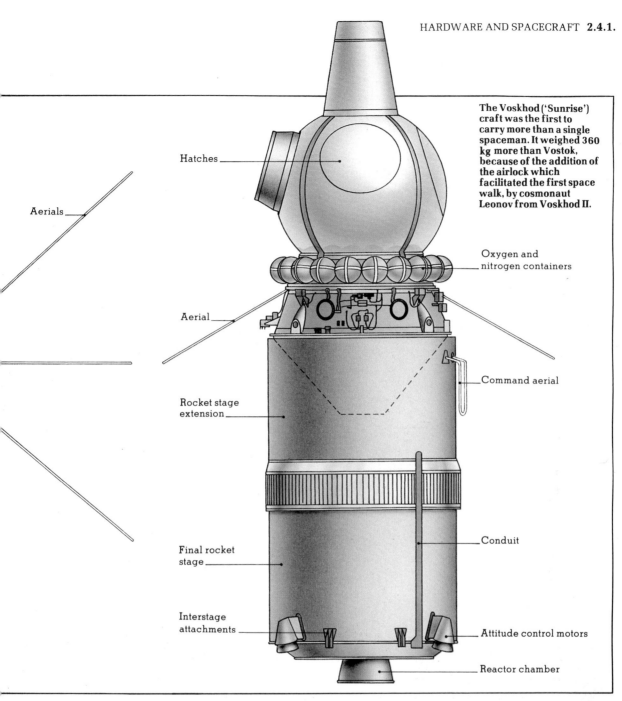

Hatches

The Voskhod ('Sunrise')
craft was the first to
carry more than a single
spaceman. It weighed 360
kg more than Vostok,
because of the addition of
the airlock which
facilitated the first space
walk, by cosmonaut
Leonov from Voskhod II.

Aerials

Oxygen and
nitrogen containers

Aerial

Command aerial

Rocket stage
extension

Conduit

Final rocket
stage

Interstage
attachments

Attitude control motors

Reactor chamber

Left: a cosmonaut lies on his
ejection seat inside the
spherical Vostok entry
capsule. A TV camera and a
porthole with an optical
orientation device are
directly in front of him.
Right: an external view of
the ship. The Vostok
programme achieved the
first manned space flight
(Vostok I), the first complete
Earth orbit (II) the
cosmonaut Titov was also
the first man to eat and sleep
in space, and to see the Sun
rise and set in one day), the
first 'formation' flight (III
and IV) and the first woman
in space (VI), Valentina
Tereshkova.

Early systems Soyuz

TWO YEARS AFTER the Voskhod flights a new spacecraft on a launcher improved by the incorporation of a more powerful second stage made its first flight. The Soyuz 1 cosmonaut was killed during re-entry, but subsequent flights had great successes, the Soyuz being used as a ferry to the Salyut space station as well as in docking operations such as that with Apollo.

In January 1969 the first exchange of passengers in space took place when Yeliseyev and Khrunov walked from Soyuz 5 to Soyuz 4. In June 1971 three cosmonauts boarded Salyut from Soyuz 11, remained there 24 days, but were killed on re-entry.

After a record 18 days in space the cosmonauts of Soyuz 9 (above) suffered severe after-effects from prolonged weightlessness.

Left: Kubasov and Leonov in the Soyuz orbital module during ASTP. Soyuz instrumentation looked relatively simple when compared with that of Apollo. Top: cosmonauts Filipchenko and Rukavishnikov after training in the Soyuz ship. Below: a Soyuz link-up. 'Soyuz' means union.

The Soyuz launcher below was approximately 50 meters tall — less than half the height of the Saturn rocket.

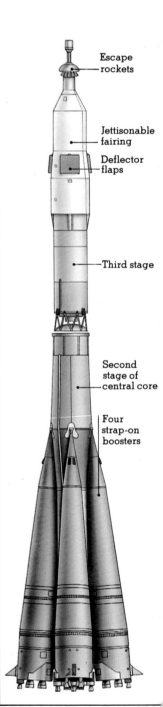

Escape rockets

Jettisonable fairing

Deflector flaps

Third stage

Second stage of central core

Four strap-on boosters

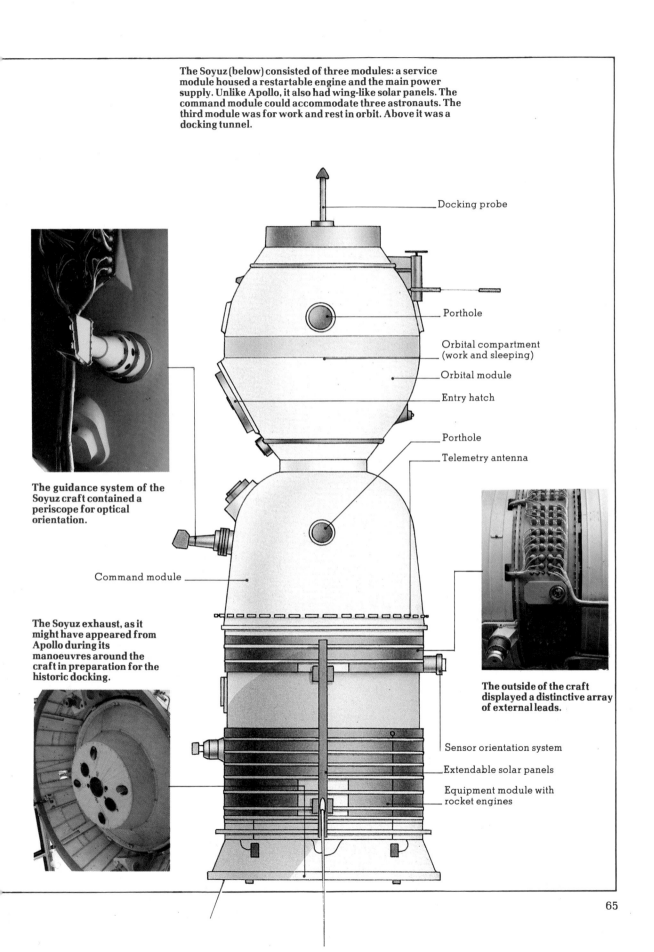

The Soyuz (below) consisted of three modules: a service module housed a restartable engine and the main power supply. Unlike Apollo, it also had wing-like solar panels. The command module could accommodate three astronauts. The third module was for work and rest in orbit. Above it was a docking tunnel.

Docking probe

Porthole

Orbital compartment (work and sleeping)

Orbital module

Entry hatch

Porthole

Telemetry antenna

Command module

Sensor orientation system

Extendable solar panels

Equipment module with rocket engines

The guidance system of the Soyuz craft contained a periscope for optical orientation.

The Soyuz exhaust, as it might have appeared from Apollo during its manoeuvres around the craft in preparation for the historic docking.

The outside of the craft displayed a distinctive array of external leads.

Early systems Apollo Soyuz Test Project

AN EARLY TENTATIVE STEP towards international co-operation that unfortunately went no further was the rendezvous and link-up of an Apollo and a Soyuz spacecraft in 1975. This historic meeting in space, called the 'Apollo Soyuz Test Project' (ASTP) was born during one of the several periods of detente between Russia and the United States during the last century.

With close co-operation between the two teams, and using a special docking module to connect the two craft (built in California by Rockwell International), the whole undertaking was successfully completed within three years of its inception.

On 17 July 1975 at 1417 hours Houston time, 2219 hours Moscow time, with the shout 'Hatch opening',

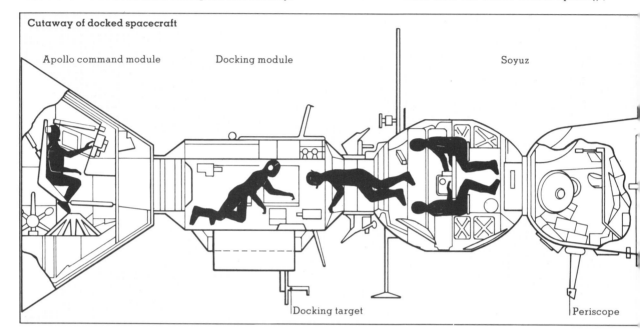

Cutaway of docked spacecraft

Apollo command module   Docking module   Soyuz

Docking target   Periscope

the two space crews docked their respective crafts together. The actual flight plans and flight manoeuvring are described on p.124, but the significance of the ASTP was its easy demonstration of the flexibility of space operations, even between two highly specific systems. One of the major differences between early Russian and American spacecraft, already discussed,

was the level of atmospheric pressure 1kg/sq cm in a typical Soyuz versus 0.35kg/sq cm in Apollo: for the rendezvous, the Russians reduced their pressure to 0.7kg/sq cm so as to shorten the transfer operations.

The Russian and American space vehicles stayed connected for nearly 47 hours and the whole mission lasted nine and a half days.

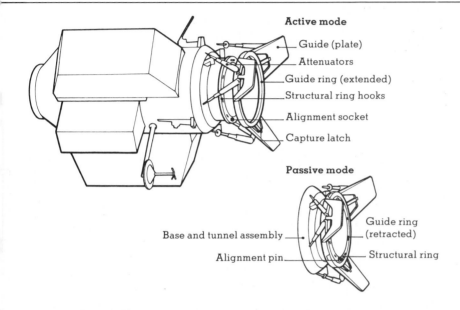

Active mode

- Guide (plate)
- Attenuators
- Guide ring (extended)
- Structural ring hooks
- Alignment socket
- Capture latch

Passive mode

- Base and tunnel assembly
- Alignment pin
- Guide ring (retracted)
- Structural ring

THE MEANS BY WHICH THEY CAME TOGETHER

Key to the operation was the Docking Module, already attached to the conical end of the Apollo Command Module. The docking was in two stages — the 'soft dock' when the craft were loosely attached to each other, and the 'hard dock' when the hatches could be opened. To start with, the guide ring was extended, and by lining up on the docking target, the two craft touched their tapered, petal-like plates together. As the plates slid against each other, spring-loaded latches locked the guide rings together. The guide rings were then retracted by attenuators and locked firmly. Concentric seals then provided a pressure seal in the tunnel area.

Far left: Apollo and Soyuz in a union which has been rather more stable and long-lasting than the hoped-for bonhomie between their respective owners.
Left: the outer mechanisms of the Docking Module described above.
Above: Soviet cosmonaut Kubasov (left) and Apollo commander Stafford in the Soviet orbital module. The considerable lengths of cable for TV and radio are partly in evidence.
Right: the Soyuz docking target.

Early systems Skylab

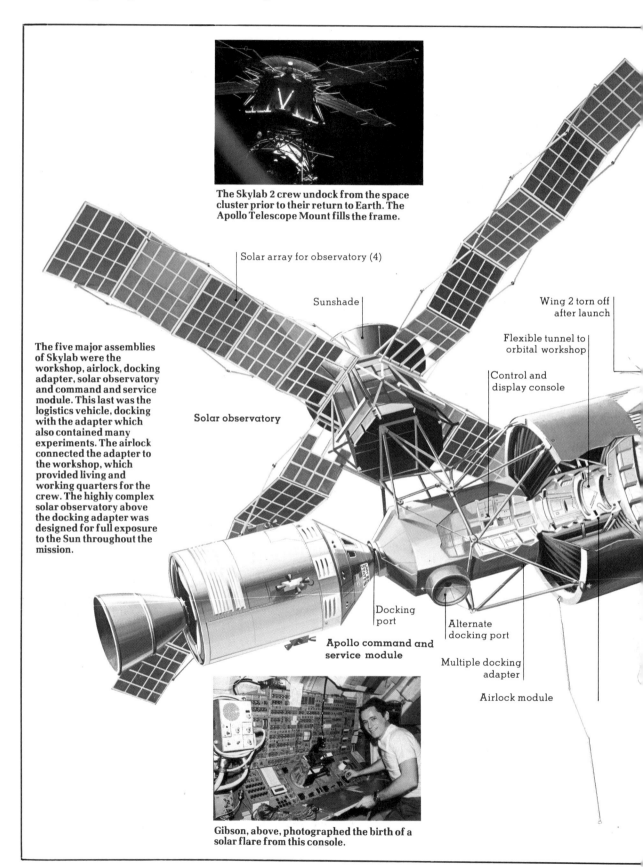

The Skylab 2 crew undock from the space cluster prior to their return to Earth. The Apollo Telescope Mount fills the frame.

Solar array for observatory (4)

Sunshade

Wing 2 torn off after launch

Flexible tunnel to orbital workshop

Control and display console

The five major assemblies of Skylab were the workshop, airlock, docking adapter, solar observatory and command and service module. This last was the logistics vehicle, docking with the adapter which also contained many experiments. The airlock connected the adapter to the workshop, which provided living and working quarters for the crew. The highly complex solar observatory above the docking adapter was designed for full exposure to the Sun throughout the mission.

Solar observatory

Docking port

Apollo command and service module

Alternate docking port

Multiple docking adapter

Airlock module

Gibson, above, photographed the birth of a solar flare from this console.

ON 14 MAY 1973, on a modified Saturn V launched from the Kennedy Space Center, Skylab began its illustrious career as the first manned space station. Although it did not last out the decade, it supported three separate crews for 171 days in space, and proved much about man's ability to live in zero-gravity.

As the intention was to use existing technology and equipment as much as possible, Skylab inevitably had a rather 'thrown-together' appearance. It was built around an empty upper stage of a Saturn rocket, fitted out with living quarters, experiment packages and a large working area. This was the workshop and, of course, fitted neatly into the launch configuration. Attached to the workshop were two 'wings' consisting of enough solar cells to power all the systems on board. A separate unit was the solar observatory, powered by its own windmill-like solar array. This observatory was positioned at right angles to the main axis of Skylab, and was attached to the specially-conceived docking adapter. The docking adapter and adjacent airlock module not only connected the workshop with the Apollo CSM that served as the arriving space transport

This view of Skylab 1 greeted the first astronauts who came to inspect the damaged craft. A micrometeoroid shield and a solar panel were lost in the launch. The partly deployed remaining panel is bottom right.

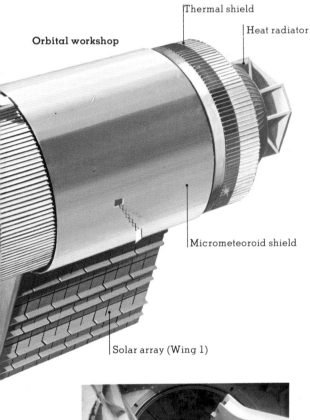

Orbital workshop

Thermal shield

Heat radiator

Micrometeoroid shield

Solar array (Wing 1)

The interior of the Multiple Docking Adapter seen from the Orbital Workshop.

Above: Skylab in orbit. Furthest from the camera are the four solar array wings of the Apollo Telescope Mount. To the right of the Orbital Workshop is the solar array wing freed by Skylab 2 crew. The corners of the 'parasol' erected to relieve the overheating in the workshop can just be seen over it.

Left: technicians at Houston's Mission Control make attitude corrections to the unmanned space station.

Early systems Skylab

for the crews, but also permitted EVA via a special hatch, and housed many of the space station controls. Together, therefore, they formed a highly-developed version of the ASTP docking module.

Despite its ultimate successes, Skylab had an inauspicious start. During launch, the micrometeoroid shield that served to protect the workshop not only from small particles but also from the heat of the Sun, tore loose, and in so doing damaged both solar 'wings', one of them so severely that it later tore away completely from the craft. With only the partial deployment of one solar array, most systems were inoperative, and even after the ground engineers' best efforts at repositioning Skylab's attitude, the cabin temperature was stabilized at 55°C. When the first crew arrived, they were able to free the deployment of the remaining solar array by EVA, and also to rig a makeshift micrometeoroid/heat shield. As a result of this and later repair work, Skylab was able to function almost normally.

After the third crew left Skylab in 1974, the space station continued to orbit in its 169 km orbit, abandoned. NASA had hoped that the first flight of the Space Shuttle would be able to carry a rocket booster to upgrade Skylab's sagging orbit, as atmospheric drag was slowing the space station so that, almost imperceptibly, it was starting to spiral Earthwards. Unfortunately, delays in the Shuttle programme prevented this, and NASA had eventually to resign itself to losing Skylab, whose orbit continued to decay until it finally broke up on re-entry.

Above: Conrad transfers equipment from the Orbital Workshop into the Airlock Module. Such work is much easier in zero-G, but objects do not lose their mass and care has to be taken that they do not cause damage by drifting into walls etc.
Left: Skylab 4 astronaut Pogue prepares to jump on the 'trash' disposal airlock hatch cover to force another bag of refuse further into the airlock. A colleague (Carr) holds two other bags, and a third floats freely.

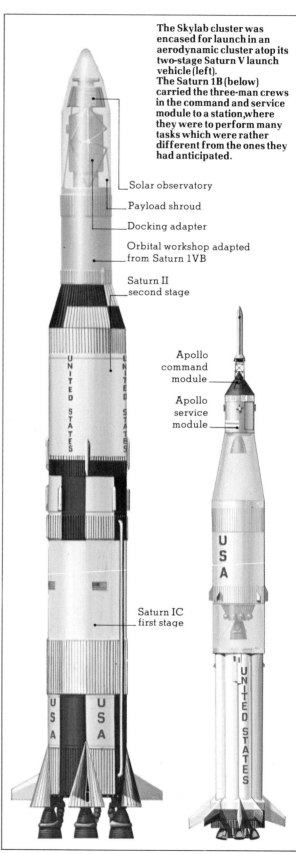

The Skylab cluster was encased for launch in an aerodynamic cluster atop its two-stage Saturn V launch vehicle (left).
The Saturn 1B (below) carried the three-man crews in the command and service module to a station, where they were to perform many tasks which were rather different from the ones they had anticipated.

Solar observatory

Payload shroud

Docking adapter

Orbital workshop adapted from Saturn 1VB

Saturn II second stage

Apollo command module

Apollo service module

Saturn IC first stage

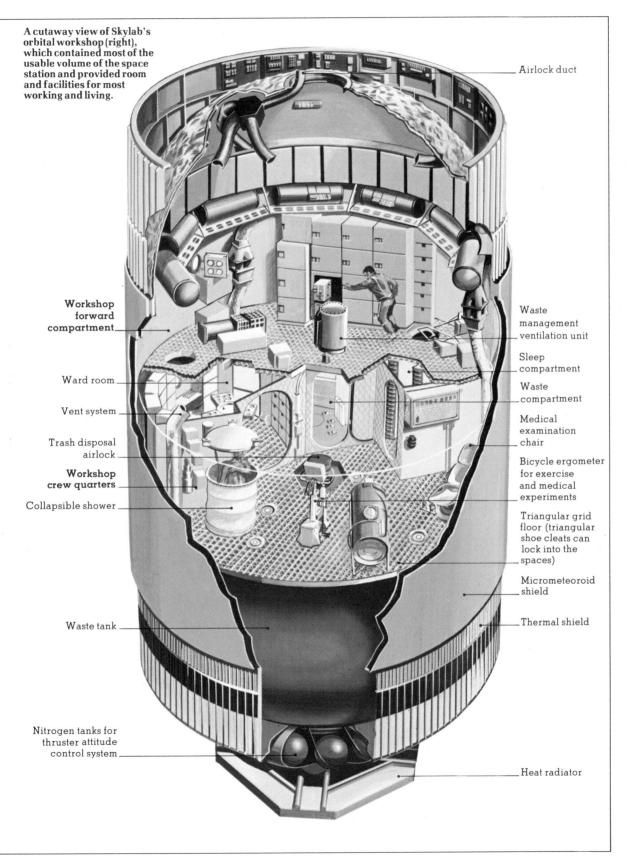

A cutaway view of Skylab's orbital workshop (right), which contained most of the usable volume of the space station and provided room and facilities for most working and living.

Airlock duct

Workshop forward compartment

Ward room

Vent system

Trash disposal airlock

Workshop crew quarters

Collapsible shower

Waste tank

Nitrogen tanks for thruster attitude control system

Waste management ventilation unit

Sleep compartment

Waste compartment

Medical examination chair

Bicycle ergometer for exercise and medical experiments

Triangular grid floor (triangular shoe cleats can lock into the spaces)

Micrometeoroid shield

Thermal shield

Heat radiator

The Shuttle Economy introduced

THE REVIVAL OF SPACE ACTIVITIES in the 1980s was centred on the American Space Shuttle. The early Space programmes had served their purpose, demonstrating Man's ability to travel in space. Nevertheless, being heavily goal-oriented, they represented something of a dead-end, and the 1970s were a hiatus in the development of spaceflight.

The Shuttle was designed to turn space operations into routine, economical missions, using the minimum of expendable hardware to bring the cost per kilo down to affordable levels. The United States began a programme of involving other nations in the Shuttle's use, emphasizing its cost-effective rather than its inspirational benefits. Later, of course, other countries acquired their own Shuttles, to the point where Shuttle lines now occupy the same status that national airlines once did. The division between heavy duty cargo vehicles and passenger vehicles has meant that the passenger Shuttle has retained much of its structure.

Configuration

Actually an aerospace vehicle, capable of manoeuvring equally well in the Earth's atmosphere and in space, the Shuttle comprises three main elements: the Orbiter, the External Tank (ET), and the Solid Rocket Boosters (SRBs). The aerodynamic Orbiter carries the crew, payload and main engines, and a typical model, such as the original Rockwell design, is much the same weight as the classic DC-9 commercial air transport — 68,000 kg without fuel. The payload bay is quite large — over 18 meters long by nearly 5 meters wide — and is designed to be able to carry a variety of cargo, from a space telescope to a completely equipped scientific laboratory. Later passenger models were built with the entire payload bay converted to seating.

The early Shuttle Orbiter carried a crew of seven and the payloads. The rocket engines of its Orbital Manoeuvring System provided control during space flight. During atmospheric flight it was controlled by the aerodynamic surfaces on the wings and by the vertical stabilizer.

The SRB's which enabled launch of the orbiter were separated at an altitude of about 45 kms and recovered from the ocean. The external tanks were ejected just before orbit and the OMS completed the insertion.

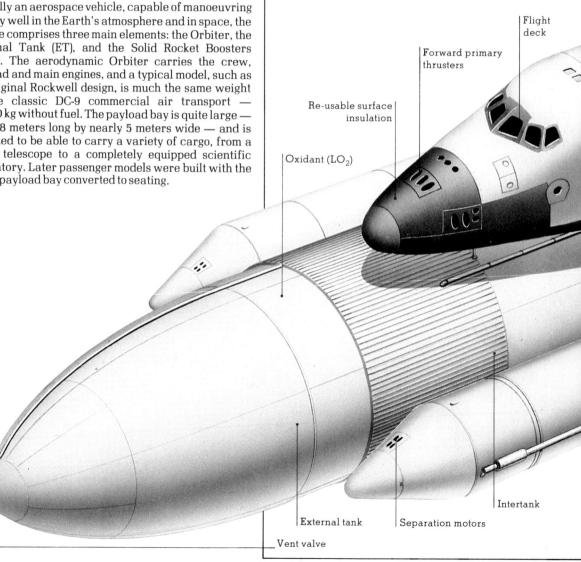

Flight deck

Forward primary thrusters

Re-usable surface insulation

Oxidant (LO$_2$)

External tank

Vent valve

Separation motors

Intertank

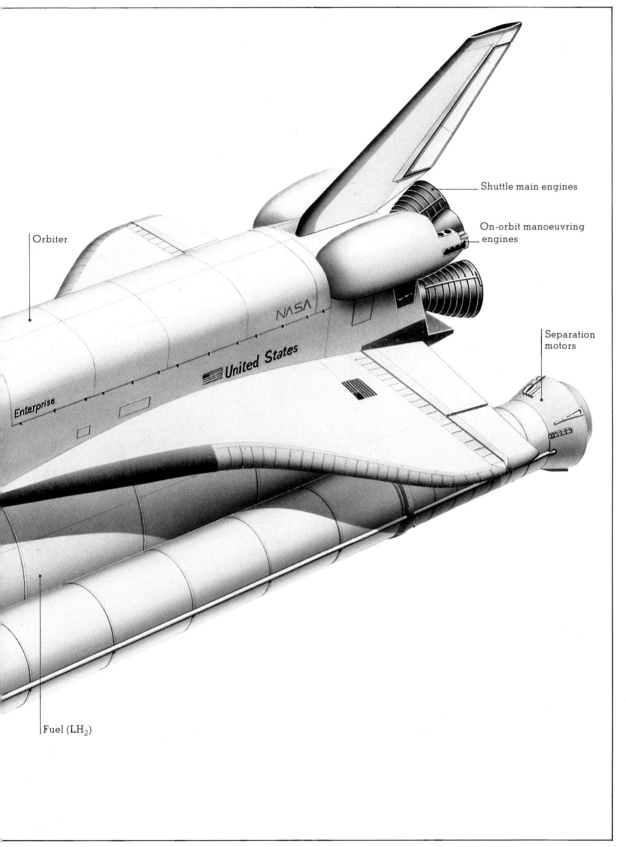

Shuttle main engines

On-orbit manoeuvring
engines

Orbiter

Separation
motors

Enterprise

United States

NASA

Fuel (LH$_2$)

Below: the parts of the early Shuttle.
Right: the original Orbiter, Enterprise, catches the early morning Sun in the Huntsville, Alabama hangar. The Marshall Space Flight Center here was responsible for the development, production and delivery of the Orbiter main engine, the solid rocket booster and the hydrogen/oxygen propellant tank.
Overleaf: a mock-up of the Orbiter at the Johnson Space Center, Houston, is equipped with a helium-filled balloon shaped to simulate a full payload. The Johnson Center was the lead Center responsible for the development, production and delivery of the Shuttle Orbiter, for programme control, systems engineering, systems integration, and for the definition of those parts of the total system that interact with other parts, such as total configuration and aerodynamic loads.

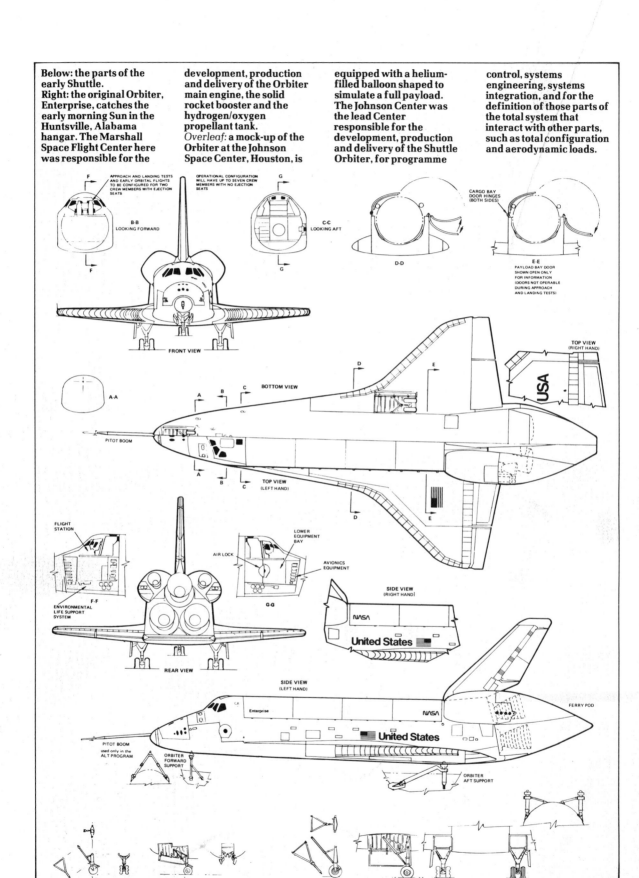

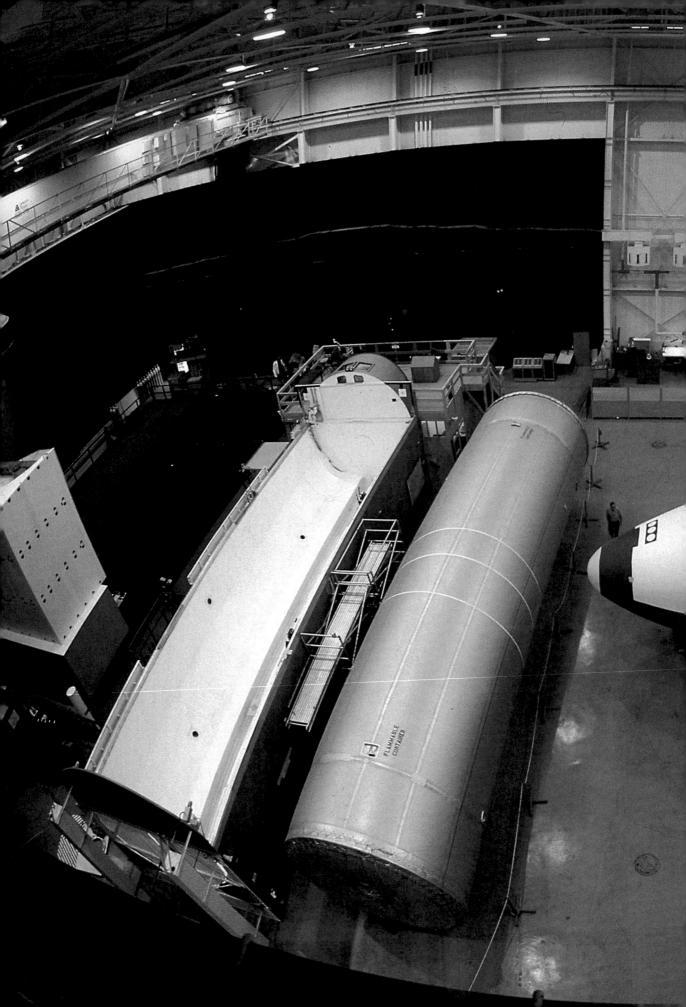

The three Space Shuttle Main Engines (SSMEs) have a thrust of 2.1 million newtons, and are among the most efficient chemical rocket engines ever designed. The fuel is stored separately, in the External Tank, which can hold 703,000kg of liquid oxygen and liquid hydrogen (in separate compartments). In the early days, the ET was the one part of the system that was not reusable, although later, constructional uses were found for the empty tanks in the space stations and colonies.

The third element of the Shuttle is the two SRBs, which, fitted to either side of the External Tank, aid launch and are recoverable.

Mission Profile

A typical Shuttle mission lasts anything from 7 to 30 days. At launch, both the boosters and the Shuttle's own engines are ignited together, and the whole configuration is launched vertically. After the fuel in the boosters is exhausted (once ignited they burn right through), they are jettisoned by explosive bolts and parachute back to Earth, where they are recovered in mid-ocean. The External Tank is jettisoned just before orbit is reached, and the Orbiter then makes its final adjustments, often involving some kind of rendezvous. Its fuel having been exhausted with the emptying of the ET, it uses a special subsystem for manoeuvring in orbit (hydrazine and nitrogen tetroxide). These propellants

Early Shuttle tests involved: transport on and launch from a Boeing jet (above); manoeuvres and rolls (left, over the Mojave Desert); and landing as if from space, with no cone on the aft end, and three dummy engines installed (below).

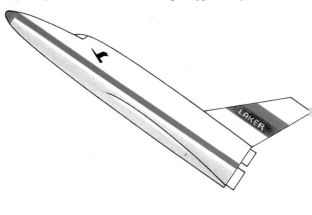

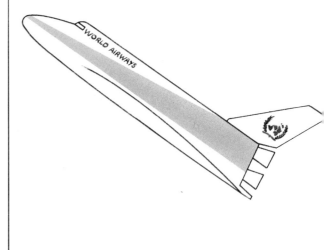

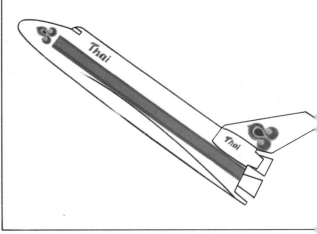

following standard airline practice in calculating passenger weights, $55,000 per person. Naturally, commercial service was not even available at the start but, realizing that air

passengers at that precise time were actually paying $25.88 per kg for a round-the-world trip (the energy expenditure is roughly the same as putting a Shuttle into orbit), Shuttle

administrators worked hard to bring orbital fares down to this level. By the first decade of the 21st century, the $50 per kg target had been reached (at constant prices), and now $40 per kg

is within sight. All this has been possible with the big increase in traffic. Over 10,000 people per year now travel by Shuttle and the rate of increase is growing all the time.

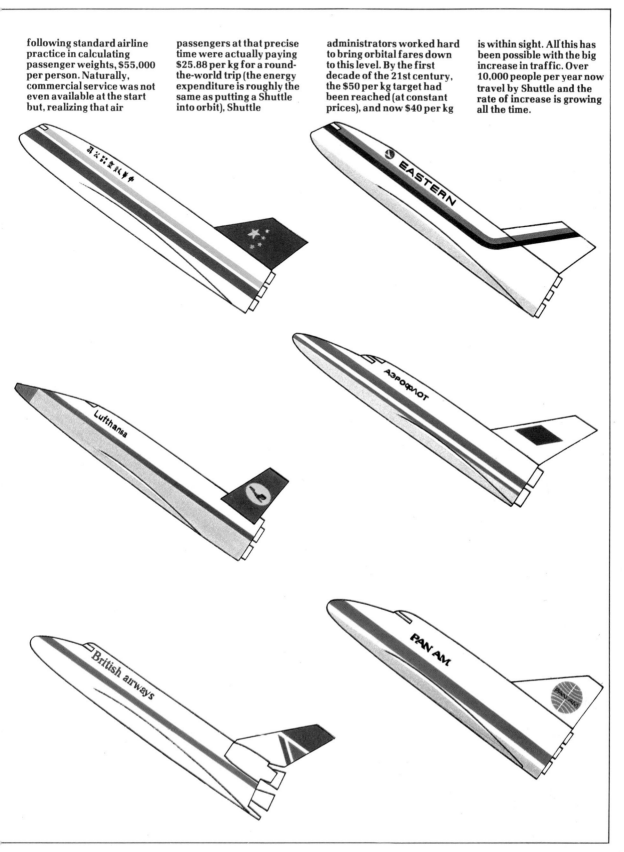

allow it some capacity for movement, but re-entry and landing are unpowered, the Orbiter's aerodynamic design allowing it to land just like an aircraft, on a runway, but at 335 km / hr. It even has a crossrange capacity (lateral movement on either side of its entry path) of more than 2000 kms. Whereas the earliest space vehicles that re-entered the Earth's atmosphere used ablation shields that eroded in order to shed heat, a special 'mosaic' insulation was developed for the Orbiter that could withstand temperatures of up to 1,260°C for up to 100 flights without needing replacement. So efficient is this insulation at shedding heat that one side can be touched with bare hands whilst the other side is red hot.

Crew

Depending on the mission, the Shuttle crew can number as many as seven: the commander, pilot, mission specialist, and up to four payload specialists. The responsibility of the mission specialist is towards the Shuttle activities that affect the payload, and on occasion this may even involve the work of a payload specialist. Payload specialists receive individual training according to the nature of the mission.

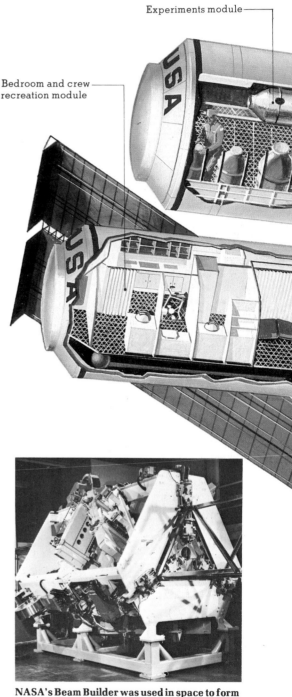

An early modular space station, based on units that could be carried in the original Orbiter payload bay.

Experiments module

Bedroom and crew recreation module

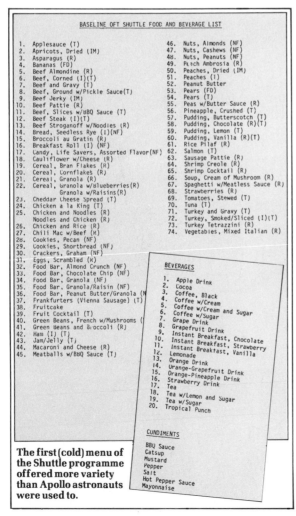

BASELINE OFT SHUTTLE FOOD AND BEVERAGE LIST

1. Applesauce (T)	46. Nuts, Almonds (NF)
2. Apricots, Dried (IM)	47. Nuts, Cashews (NF)
3. Asparagus (R)	48. Nuts, Peanuts (NF)
4. Bananas (FD)	49. Peach Ambrosia (R)
5. Beef Almondine (R)	50. Peaches, Dried (IM)
6. Beef, Corned (I)(T)	51. Peaches (I)
7. Beef and Gravy (T)	52. Peanut Butter
8. Beef, Ground w/Pickle Sauce(T)	53. Pears (FD)
9. Beef Jerky (IM)	54. Pears (T)
10. Beef Pattie (R)	55. Peas w/Butter Sauce (R)
11. Beef, Slices w/BBQ Sauce (T)	56. Pineapple, Crushed (T)
12. Beef Steak (I)(T)	57. Pudding, Butterscotch (T)
13. Beef Stroganoff w/Noodles (R)	58. Pudding, Chocolate (R)(T)
14. Bread, Seedless Rye (I)(NF)	59. Pudding, Lemon (T)
15. Broccoli au Gratin (R)	60. Pudding, Vanilla (R)(T)
16. Breakfast Roll (I) (NF)	61. Rice Pilaf (R)
17. Candy, Life Savers, Assorted Flavor(NF)	62. Salmon (T)
18. Cauliflower w/Cheese (R)	63. Sausage Pattie (R)
19. Cereal, Bran Flakes (R)	64. Shrimp Creole (R)
20. Cereal, Cornflakes (R)	65. Shrimp Cocktail (R)
21. Cereal, Granola (R)	66. Soup, Cream of Mushroom (R)
22. Cereal, Granola w/blueberries(R)	67. Spaghetti w/Meatless Sauce (R)
Granola w/Raisins(R)	68. Strawberries (R)
23. Cheddar Cheese Spread (T)	69. Tomatoes, Stewed (T)
24. Chicken a la King (T)	70. Tuna (T)
25. Chicken and Noodles (R)	71. Turkey and Gravy (T)
Noodles and Chicken (R)	72. Turkey, Smoked/Sliced (I)(T)
26. Chicken and Rice (R)	73. Turkey Tetrazzini (R)
27. Chili Mac w/Beef (R)	74. Vegetables, Mixed Italian (R)
28. Cookies, Pecan (NF)	
29. Cookies, Shortbread (NF)	
30. Crackers, Graham (NF)	
31. Eggs, Scrambled (R)	
32. Food Bar, Almond Crunch (NF)	
33. Food Bar, Chocolate Chip (NF)	
34. Food Bar, Granola (NF)	
35. Food Bar, Granola/Raisin (NF)	
36. Food Bar, Peanut Butter/Granola (N	
37. Frankfurters (Vienna Sausage) (T)	
38. Fruitcake	
39. Fruit Cocktail (T)	
40. Green Beans, French w/Mushrooms (	
41. Green Beans and Broccoli (R)	
42. Ham (I) (T)	
43. Jam/Jelly (T)	
44. Macaroni and Cheese (R)	
45. Meatballs w/BBQ Sauce (T)	

BEVERAGES

1. Apple Drink
2. Cocoa
3. Coffee, Black
4. Coffee w/Cream
5. Coffee w/Cream and Sugar
6. Coffee w/Sugar
7. Grape Drink
8. Grapefruit Drink
9. Instant Breakfast
10. Instant Breakfast, Chocolate
11. Instant Breakfast, Strawberry
12. Instant Breakfast, Vanilla
13. Lemonade
14. Orange Drink
15. Orange-Grapefruit Drink
16. Orange-Pineapple Drink
17. Strawberry Drink
18. Tea
19. Tea w/Lemon and Sugar
20. Tea w/Sugar
    Tropical Punch

CONDIMENTS

BBQ Sauce
Catsup
Mustard
Pepper
Salt
Hot Pepper Sauce
Mayonnaise

The first (cold) menu of the Shuttle programme offered more variety than Apollo astronauts were used to.

NASA's Beam Builder was used in space to form construction beams for large structures. The aluminium beams could be very light, because of the absence of stress in orbit.

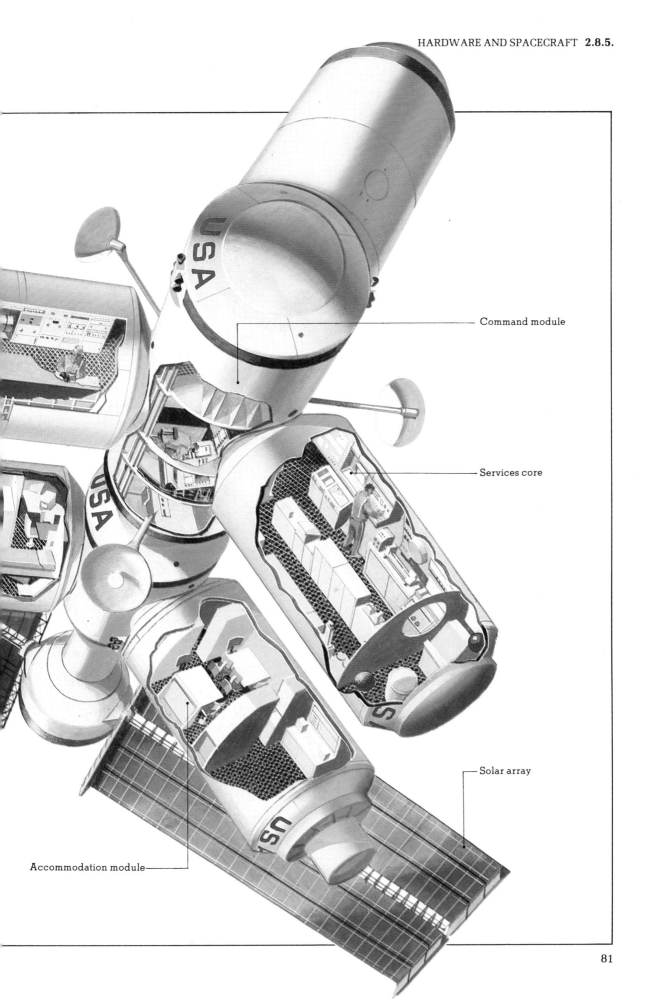

Command module

Services core

Solar array

Accommodation module

The Shuttle Derivatives

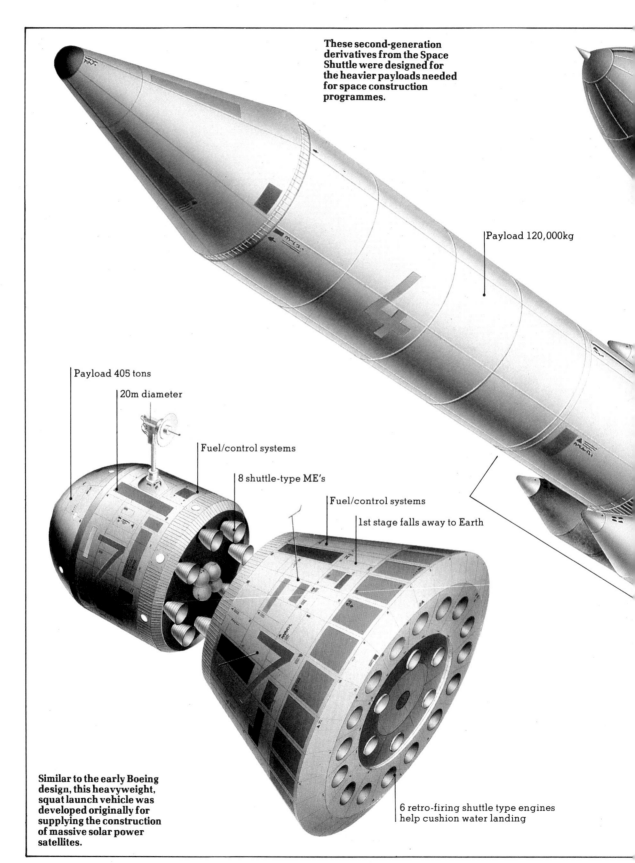

These second-generation derivatives from the Space Shuttle were designed for the heavier payloads needed for space construction programmes.

Payload 120,000kg

Payload 405 tons

20m diameter

Fuel/control systems

8 shuttle-type ME's

Fuel/control systems

1st stage falls away to Earth

Similar to the early Boeing design, this heavyweight, squat launch vehicle was developed originally for supplying the construction of massive solar power satellites.

6 retro-firing shuttle type engines help cushion water landing

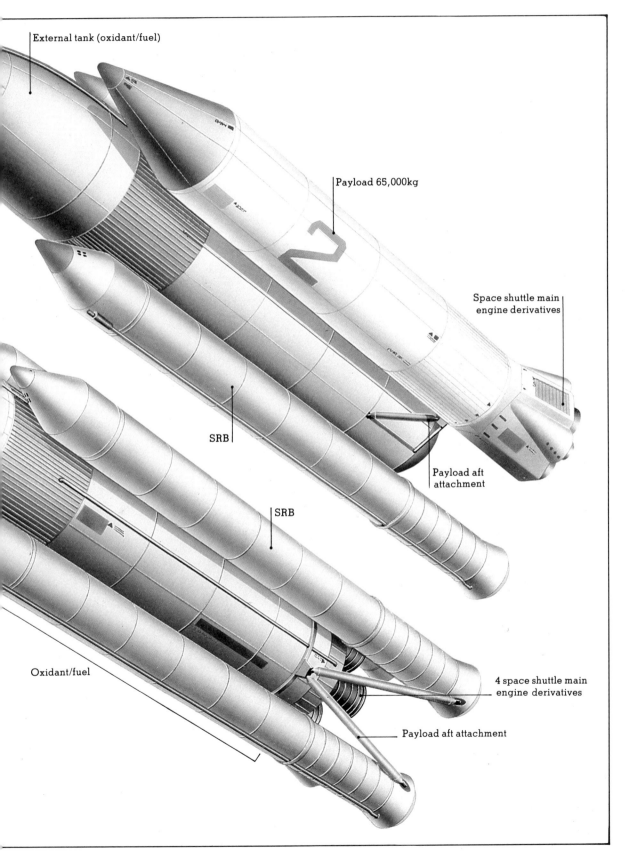

External tank (oxidant/fuel)

Payload 65,000kg

Space shuttle main engine derivatives

SRB

Payload aft attachment

SRB

Oxidant/fuel

4 space shuttle main engine derivatives

Payload aft attachment

Orbital Transfer Vehicles The space tugs

FOR REASONS OF FUEL ECONOMY, Earth-to-orbit shuttles of all designs deliver payloads only as far as a Low Earth Orbit (LEO), that is, to a maximum altitude of 500 km. Most payloads, however, need to go higher, whether to the stable Geosynchronous Earth Orbit at 35,900 km, Lagrange orbits, the Moon, or a deep-space launching. Transfer to higher orbits is the work of the space tugs, known as Orbital Transfer Vehicles (OTVs) — the workhorses of space. Modular in design, the OTVs can be adapted to a number of different tasks simply by bolting on additional structures. The basic space tug is a simple, no-frills chemical rocket with an internal payload capacity of 820,000 kg, integral fuel tanks and a remote guidance system that is usually controlled from the shuttle and the receiving station. Ferrying materials around the Earth-Moon system is such a basic task that there is little room for technological improvement, and indeed there has been little since OTV's were introduced in 1992.

A crew module can be bolted on for manned operations, and additional small manipulation modules permit reasonably delicate construction or loading work. These modules are equipped with mechanical arms and are designed for one-man operation. Umbilical extensions to the crew module allow up to three manipulators to work at moderate distances from the mother craft, using small hydrazine-powered thrusters (these must be specially fitted).

Additional payloads can be attached to the tug in many combinations, either in modular tanks or on custom rigs if the freight is large or of an awkward shape. There is, in fact, almost no limit to the configurations of a loaded OTV. In the 1990s the tug was adapted for lunar landings. This was done quite simply, by fitting landing gear and a landing guidance system; the rocket, derived from the highly efficient Space Shuttle main engine, is quite up to the task. However, the fast pace of early lunar base construction made it more practicable to build a fleet of Lunar Lander OTVs (see p.102) with integrated crew compartments, and these are still in service (with modifications).

Of course, for certain missions orbital transfer is sometimes conducted by other craft, and both nuclear-thermal rockets and solar-electric ion rockets are occasionally seen on freight duty.

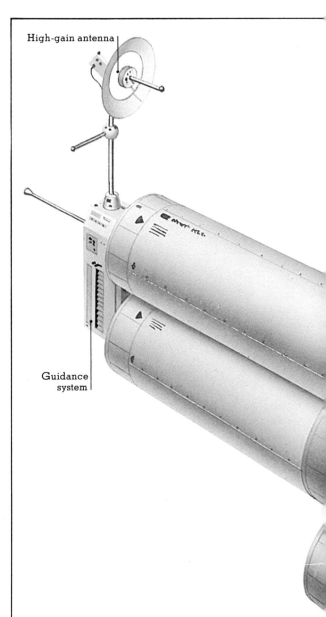

High-gain antenna

Guidance system

Left: a familiar sight during the peak period of lunar base construction. The tug manipulator modules contributed to the effect of tremendous activity which rather surprised early visitors.
Right: a tug in its loaded configuration. As with the old packhorse, loads tend to be added to it almost ad infinitum but, unlike its ancient predecessor, it tends not to tire.
Top right: the Space Shuttle Main Engine, in its later role as rocket for the space tug.

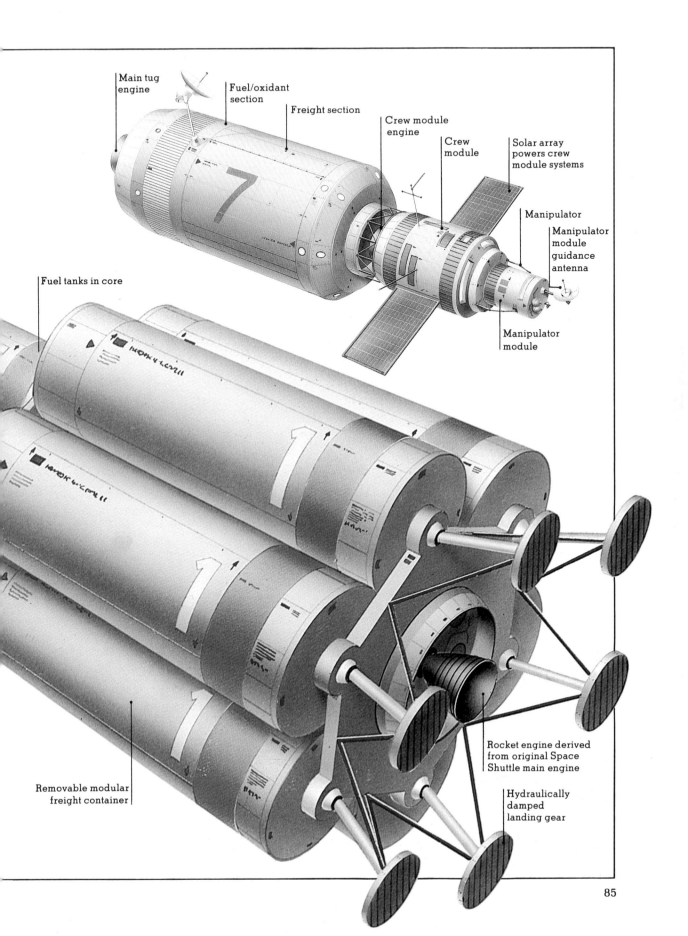

Main tug engine

Fuel/oxidant section

Freight section

Crew module engine

Crew module

Solar array powers crew module systems

Manipulator

Manipulator module guidance antenna

Manipulator module

Fuel tanks in core

Rocket engine derived from original Space Shuttle main engine

Hydraulically damped landing gear

Removable modular freight container

The ion rocket Low thrust for long missions

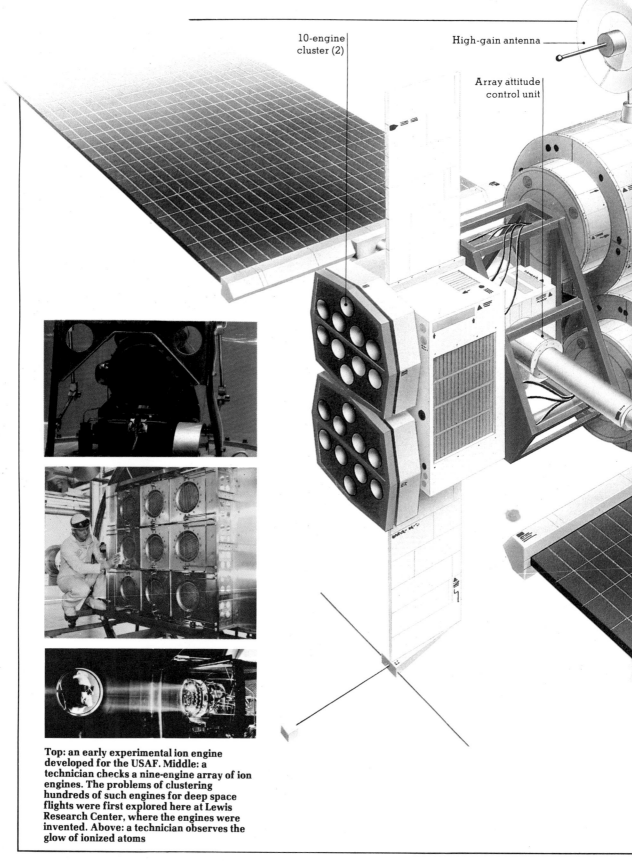

10-engine cluster (2)

High-gain antenna

Array attitude control unit

Top: an early experimental ion engine developed for the USAF. Middle: a technician checks a nine-engine array of ion engines. The problems of clustering hundreds of such engines for deep space flights were first explored here at Lewis Research Center, where the engines were invented. Above: a technician observes the glow of ionized atoms

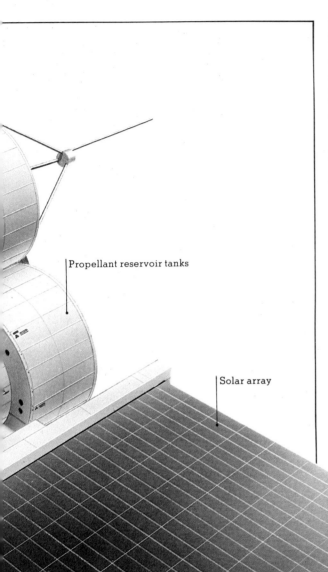

Propellant reservoir tanks

Solar array

USING ELECTRIC PROPULSION, ion rockets offer an ideal transportation system for unmanned missions over long distances where time is not an important factor. The propellant, which can be dense and therefore occupy little space, is ionized and electrically exhausted to provide a low thrust at a very high velocity. Most interesting of all, perhaps, is that the electrical energy needed to ionize the fuel comes free, from the Sun.

Solar energy is converted into electrical energy by large solar panels giving ion rockets their characteristic appearance. The most efficient, and therefore most common propellants, are mercury and cesium gases. The atoms of either of these gases are ionized (stripped of their electrons) in the discharge chamber by being bombarded with electrons. At this point the ions and the free electrons form a plasma, and an electric field sucks out the ions at great speed through a fine screen grid. As this ion beam has one charge, it must be neutralized by the injection of an equal number of electrons.

Although the thrust is extremely low, giving the spacecraft a slow send-off, the velocity of the ion beam is very high, and with continuous operation eventually takes the spacecraft up to a respectable velocity. It is for this reason that ion rockets are so well suited to long missions.

Auxiliary thrusters, for station-keeping and mid-course corrections, can also be electrically powered, and over long missions save a considerable amount of weight over cold-gas jets. On shorter missions, however, the saving is less, and it is more normal to use hydrazine thrusters for station-keeping.

Nuclear-thermal rocket Power to rely on

THE FIRST NUCLEAR ROCKET design, tested in the 1960s, is one in which hydrogen fuel is heated to a very high temperature by a solid core reactor. In principle this is the most conventional use of nuclear power, not as exotic as the new generation of nuclear pulse motors, but very serviceable for interplanetary use.

The original model, NERVA (Nuclear Engine for Rocket Vehicle Application) was first tested at Jackass Flats in Nevada in 1969, but budgetary cut-backs later shelved the whole project. Since revived, principally for Mars and Jupiter missions, the nuclear-thermal rocket has proved a reliable workhorse.

A turbo-pump delivers the propellant, liquid hydrogen (LH), from the forward storage tanks at engine operating pressure. Before being heated by the reactor, the hydrogen is first passed around the nozzle to cool it, and then up into the reactor core. Now extremely hot, the hydrogen passes through the exhaust nozzle,

The components of the NERVA nuclear rocket engine (right) are, from the top: base of propellant tank; spherical bottles for gases; liquid hydrogen turbo-pump housing; roll control nozzles; reactor control-rod actuators; thrust chamber liquid hydrogen coolant line to hydrogen-cooled nozzle. The engine works by pumping super-cold liquid hydrogen through the nuclear reactor to be super-heated so that it expands violently out of the nozzle. The vehicle (far right), having just left Earth orbit on its journey to Mars, would have just brought its nuclear rockets into play.

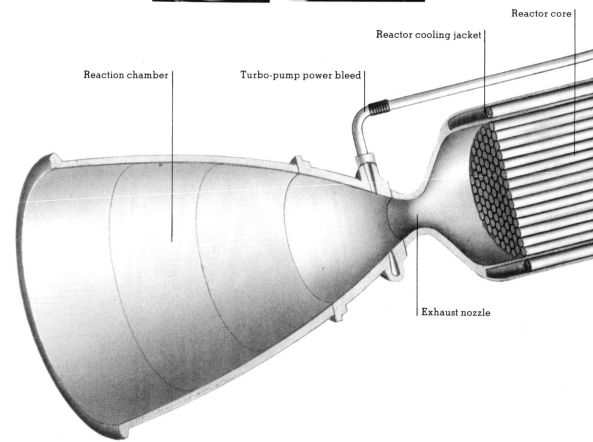

Reaction chamber

Turbo-pump power bleed

Reactor cooling jacket

Reactor core

Exhaust nozzle

expanding and accelerating to produce thrust. A small amount is bled off to power the turbo-pump.

A highly-efficient engine, the nuclear-thermal rocket's performance (specific impulse) is more than twice that of ordinary chemical rockets. Of course to achieve this it needs more power, but the intense concentration of energy in the small nuclear package easily solves this problem. In addition, nuclear energy is easily 'stored', being quite stable.

A small nuclear-thermal interplanetary rocket (similar to the early NERVA) can make a round trip to Mars in 600 days on five propellant tanks, or, for example, can deliver 1,800kg to Neptune in six and a half years. If a gravity-assist from Jupiter is engineered through appropriate timing of launch etc., then this figure can be reduced to 4.8 years. Large rockets have up to five times NERVA's thrust of 34 tonnes, and can deliver much heavier payloads.

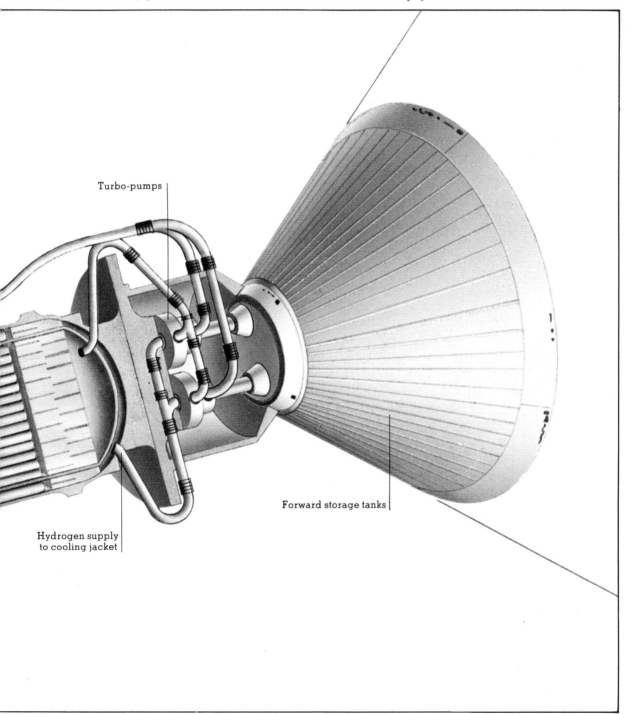

Turbo-pumps

Hydrogen supply
to cooling jacket

Forward storage tanks

The mass driver Mover of asteroids

SOMETIMES KNOWN as the magnetic levitation accelerator, the mass-driver engine employs a principle first used economically in 1972 on a Japanese railroad. Under the right conditions, an electromagnetic track can not only levitate a vehicle with its magnetic field (thereby reducing friction) but can also accelerate it along. With sophisticated modifications, this is the principle of the mass-driver, now used on the Moon to launch mined rock for construction use at the Lagrange Colonies, and also for the transport of whole asteroids from the Belt inwards to more convenient mining locations in the Inner System.

In a little more detail, the mass-driver as an asteroid-mover works like this: electrical energy, which as in the case of ion rockets can come from the Sun via large solar arrays, drives magnetic impulses along a drive

coil — a tubular 'track' made of superconducting materials. The 'vehicles' that use this track are a kind of bucket, consisting principally of two ring-shaped coils and a container. These buckets carry pulverized rock mined from the asteroid and used purely as reaction mass. In accordance with Isaac Newton's Third Law of Motion, accelerating reaction mass away from the spacecraft provides the thrust that drives the craft forward. Here, the asteroid is the spacecraft and to move forward it must expel part of its own mass.

In 1977 MIT students put together an Earthbound demonstration mass-driver in the laboratory, capable of accelerating to 33G in 0.11 sec. Although the magnets cannot levitate the buckets when they are at rest, they function even at the low speed of 32 km/hr. The successive triggering of a series of coils functions as a

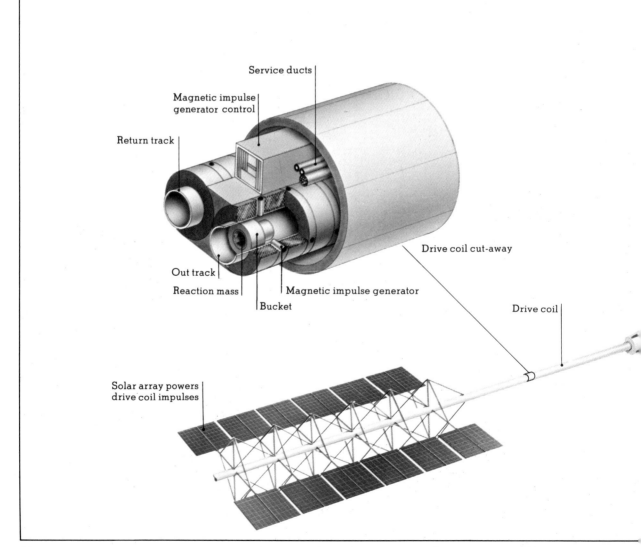

For prospective Belt miners: the shape, size and make-up of their capital investment. Top left, a cross section of the drive-coil.

Service ducts

Magnetic impulse generator control

Return track

Out track

Reaction mass

Bucket

Magnetic impulse generator

Drive coil cut-away

Drive coil

Solar array powers drive coil impulses

linear motor, which, in a vacuum, can be highly effective. In use in space, the buckets, which return along a loop system, follow each other rapidly, in the order of two every second. As each bucket returns to the asteroid end of the accelerator, it slows momentarily, passing through a rapid cooling state to improve its performance. A measured quantity of pulverized rock is injected into the bucket, which is then quickly accelerated along the track by the coils. At the far end, often kilometers away from the asteroid, the bucket releases its payload, sharply decelerating so that the load flies off into space. The bucket turns 180° along a loop, and the cycle begins over again. Initial acceleration of the asteroid is slow, and journeys are measured in years.

In setting up an asteroid for propulsion in this way, the mining assembly is first towed into position. Anchor 'spikes' are inserted manually into charge-blown holes. Guy-lines from the mining assembly are attached to these 'spikes' and winched tight, so attaching the whole assembly to the asteroid. Then small drilling bits on the end of long arms bite into the rock. Spring-loaded, backward-pointing blades surrounding each bit are released next, anchoring the whole structure firmly enough to start mining operations. The central drilling assembly is massive and adjustable, passing the mined rubble backwards like a mole. The rubble is then further processed ready for injection into the mass-driver's buckets. The drilling assembly adjusts forward automatically as it bites further into the asteroid. At the limit of its extension, it is withdrawn, and everything is moved to a new site on the asteroid's surface.

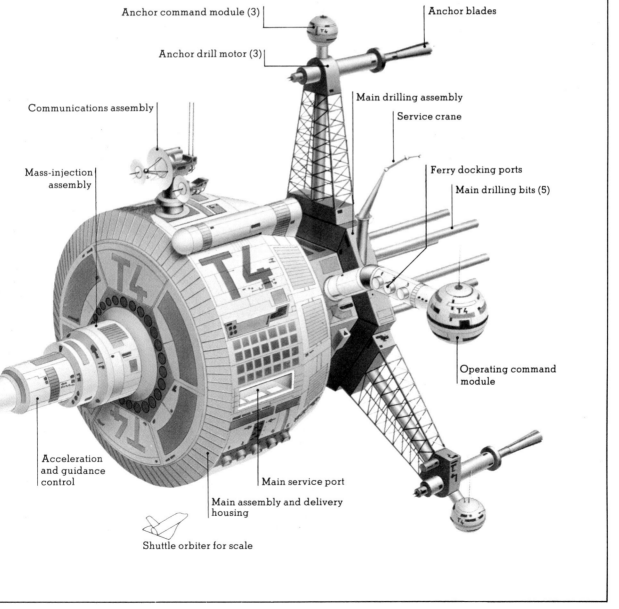

Anchor command module (3)

Anchor blades

Anchor drill motor (3)

Main drilling assembly

Service crane

Communications assembly

Ferry docking ports

Mass-injection assembly

Main drilling bits (5)

Operating command module

Acceleration and guidance control

Main service port

Main assembly and delivery housing

Shuttle orbiter for scale

Solar sails Catching the wind

UNQUESTIONABLY THE MOST ELEGANT method of space travel, solar sailing is now undergoing a popular revival among do-it-yourself enthusiasts in the Lagrange colonies. Although successfully tested in the 1980s by the Jet Propulsion Laboratory, and used a number of times, the sail was largely ignored for nearly a century in favour of other systems. Now, however, its basic disadvantage — launching and deploying from Earth — no longer applies, and sails of many different designs are now constructed in the marinas adjacent to the colonies.

The concept is an engaging one: using the simple pressure of sunlight in much the same way as a yacht uses the wind. There have not been any significant improvements in design since the early days, largely because the sail has never been taken up commercially. Nevertheless, respectable performances can be achieved using space-manufactured aluminium foil 1/100,000 mm thick and fine-gauge sail wire.

The basic hurdle has always been mass. The pressure of sunlight in the vicinity of the Earth is very feeble — exerting a force on a square kilometer of sail only sufficient to accelerate one kilogram at half a g. With a total mass of sail and payload of 5000 kg, this acceleration is reduced to only 0.0001 G (a 1/10,000). Surprisingly, though, this builds up to a healthy velocity with time because, of course, the sunlight is a *constant* pressure. So, after nearly two weeks the sail would have been accelerated to one kilometer per second.

These figures can be improved on by reducing the payload to a minimum and using a large sail.

The sail shown here is of the standard hexagonal shape and measures approximately 10 kilometers across.

SAILING IN SPACE

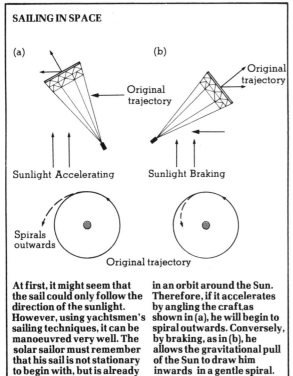

(a)

(b)

Original trajectory

Original trajectory

Sunlight Accelerating

Sunlight Braking

Spirals outwards

Original trajectory

At first, it might seem that the sail could only follow the direction of the sunlight. However, using yachtsmen's sailing techniques, it can be manoeuvred very well. The solar sailor must remember that his sail is not stationary to begin with, but is already in an orbit around the Sun. Therefore, if it accelerates by angling the craft, as shown in (a), he will begin to spiral outwards. Conversely, by braking, as in (b), he allows the gravitational pull of the Sun to draw him inwards in a gentle spiral.

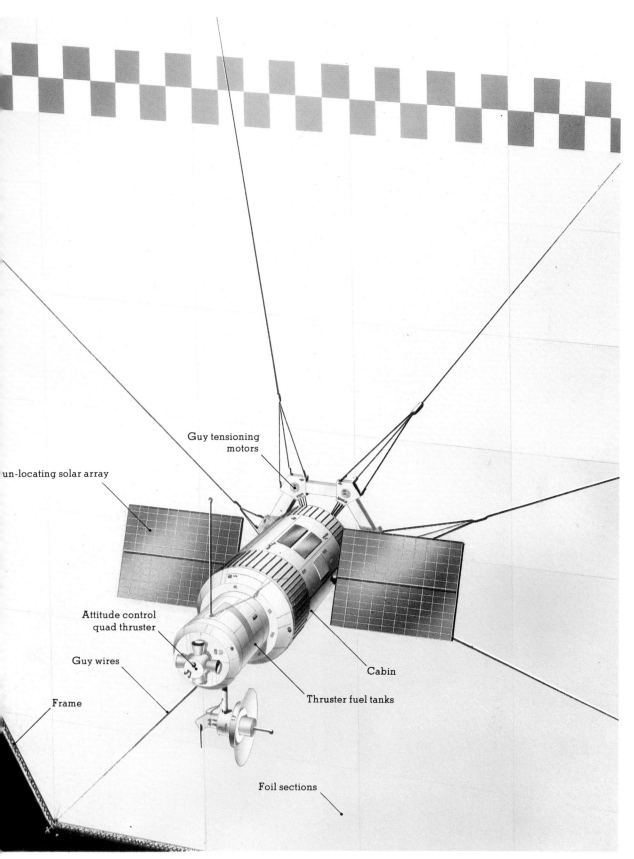

Guy tensioning
motors

un-locating solar array

Attitude control
quad thruster

Guy wires

Frame

Cabin

Thruster fuel tanks

Foil sections

Nuclear pulse rockets Stars within reach

THE FIRST fully developed starship propulsion system is the nuclear pulse engine, made famous by the original interstellar probe, Daedalus. This exciting advance in rocketry will surely open the way to manned galactic travel, making possible velocities far in advance of chemical, electric and nuclear-thermal systems.

Whilst nuclear explosions for propulsion were examined as early as the 1950s, it was only the development of fusion technology that made starships possible. Exploded at a rate of 250 per second, the small deuterium-helium 3 'bombs' accelerated Daedalus over four years to 12.2 per cent of the speed of light.

Following a long testing programme directed from the Callisto control centre, the starship left Jupiter Space for Barnard's star and its planetary system, 5.9 light years away, on 1 July 2049.

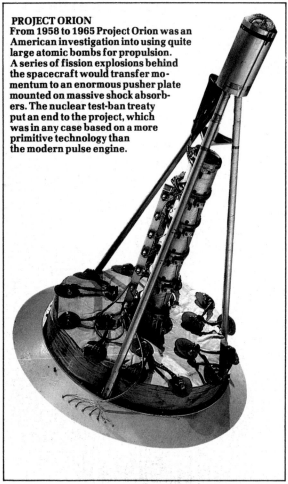

PROJECT ORION
From 1958 to 1965 Project Orion was an American investigation into using quite large atomic bombs for propulsion. A series of fission explosions behind the spacecraft would transfer momentum to an enormous pusher plate mounted on massive shock absorbers. The nuclear test-ban treaty put an end to the project, which was in any case based on a more primitive technology than the modern pulse engine.

The second stage of the starship Daedalus is photographed here during the lengthy testing operations in a manoeuvre in Uranus space.

PROPULSION OF THE STARSHIP

Fuel pellets stored cryogenically in the spherical tanks are injected one by one into the reaction chamber by an electromagnetic gun. Just as each pellet reaches the target point, a ring of high-powered electron beams hits it with sufficient energy to generate a thermonuclear fusion reaction. The magnetic field inside the chamber is compressed, briefly absorbs the force of the explosion, and then pushes the plasma outwards. An induction loop at the exit picks off some of the energy to re-supply the electron beam generators. Using a fuel mixture of deuterium and helium-3 mined from Jupiter's atmosphere by hot-air balloon extraction plants (see p. 182), 30 billion pellets are used on the voyage, each one enclosed in a thin superconducting shell. The advantage of this fuel is that the reaction products can be controlled by the magnetic field, and there are relatively few neutrons, so reducing the radiation.

Fuel pellet of D_3He

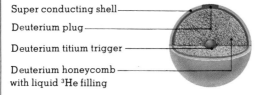

Super conducting shell

Deuterium plug

Deuterium titium trigger

Deuterium honeycomb with liquid ^3He filling

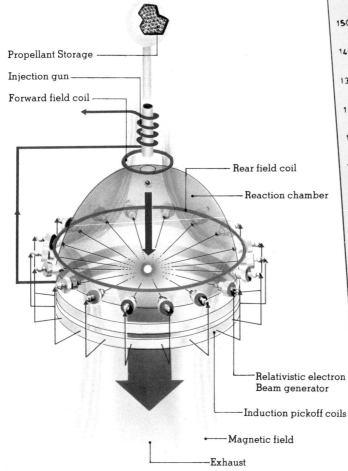

Propellant Storage

Injection gun

Forward field coil

Rear field coil

Reaction chamber

Relativistic electron Beam generator

Induction pickoff coils

Magnetic field

Exhaust

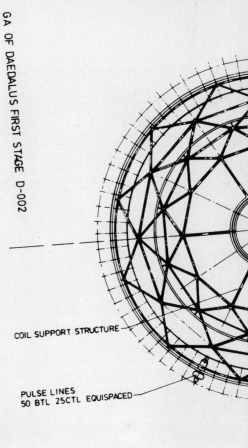

COIL SUPPORT STRUCTURE

PULSE LINES
50 BTL 25CTL EQUISPACED

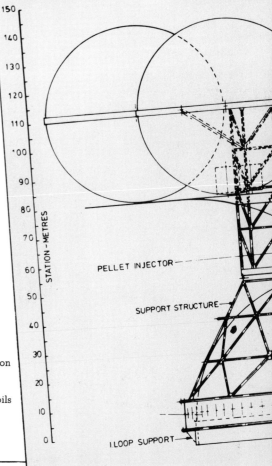

150
140
130
120
110
100
90
80
70
60
50
40
30
20
10
0

STATION - METRES

PELLET INJECTOR

SUPPORT STRUCTURE

I LOOP SUPPORT

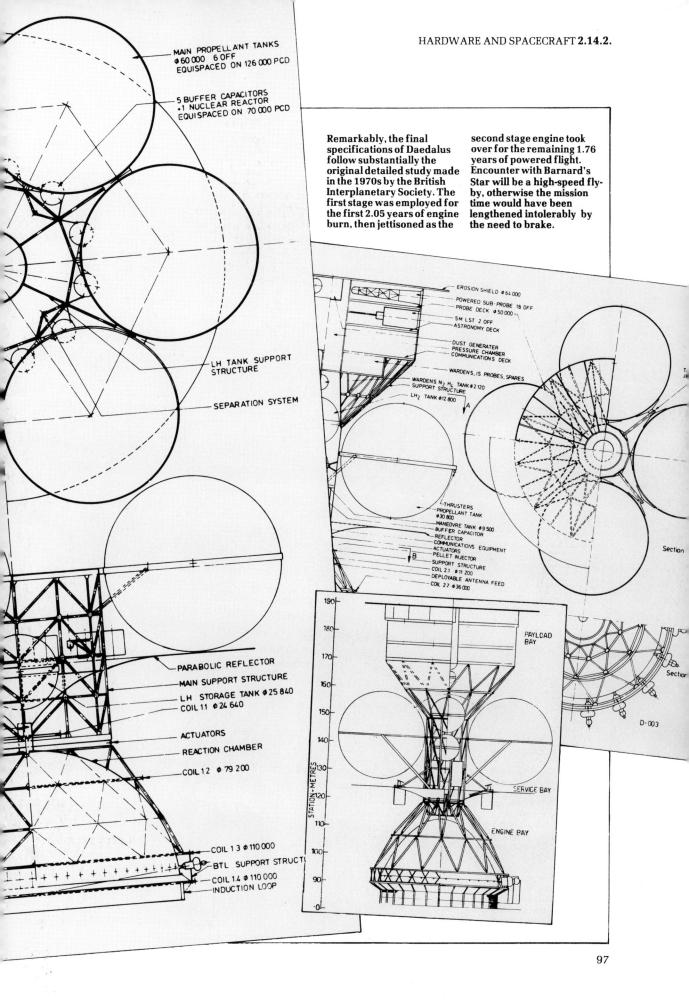

MAIN PROPELLANT TANKS
φ 60 000   6 OFF
EQUISPACED ON 126 000 PCD

5 BUFFER CAPACITORS
+1 NUCLEAR REACTOR
EQUISPACED ON 70 000 PCD

LH TANK SUPPORT
STRUCTURE

SEPARATION SYSTEM

Remarkably, the final specifications of Daedalus follow substantially the original detailed study made in the 1970s by the British Interplanetary Society. The first stage was employed for the first 2.05 years of engine burn, then jettisoned as the second stage engine took over for the remaining 1.76 years of powered flight. Encounter with Barnard's Star will be a high-speed fly-by, otherwise the mission time would have been lengthened intolerably by the need to brake.

EROSION SHIELD φ 64 000
POWERED SUB-PROBE 18 OFF
PROBE DECK φ 50 000
5M LST 2 OFF
ASTRONOMY DECK
DUST GENERATOR
PRESSURE CHAMBER
COMMUNICATIONS DECK
WARDENS, 15 PROBES, SPARES
WARDENS N$_2$ H$_2$ TANK φ 2 120
SUPPORT STRUCTURE
LH$_2$ TANK φ 12 800

THRUSTERS
PROPELLANT TANK
φ 30 800
MANEOVRE TANK φ 9 500
BUFFER CAPACITOR
REFLECTOR
COMMUNICATIONS EQUIPMENT
ACTUATORS
PELLET INJECTOR
SUPPORT STRUCTURE
COIL 2.1 φ 11 200
DEPLOYABLE ANTENNA FEED
COIL 2.2 φ 36 000

Section

Section

D-003

PARABOLIC REFLECTOR
MAIN SUPPORT STRUCTURE
LH STORAGE TANK φ 25 840
COIL 1.1 φ 24 640

ACTUATORS

REACTION CHAMBER

COIL 1.2 φ 79 200

COIL 1.3 φ 110 000
BTL SUPPORT STRUCT
COIL 1.4 φ 110 000
INDUCTION LOOP

STATION-METRES

PAYLOAD BAY

SERVICE BAY

ENGINE BAY

The ramjet Interstellar breakthrough

CURRENTLY UNDER TEST at Callisto Base, the ramjet offers the greatest promise for manned interstellar travel. Now that the galactic hydrogen fields have been mapped with reasonable certainty, average scoop sizes and theoretical operational speeds are now known.

In principle, the ramjet dispenses with the great bugbear of rocketry. — the need to carry large quantities of propellant. All conventional rockets waste much of their energy accelerating propellant reserves

and the heavy mass-ratios are a serious limit to missions. The rocket engineer's dream has been to collect fuel on the way, and now this may be possible.

The interstellar fuel is hydrogen, which in dense galactic fields occurs at up to 100 atoms per cc. Although this may seem an extremely low density, project engineers have calculated that a collecting area of 10,000 sq km will be sufficient to maintain a healthy acceleration for a 1000 tonne ship. The ramscoop is magnetic; the powerful field generators distributed

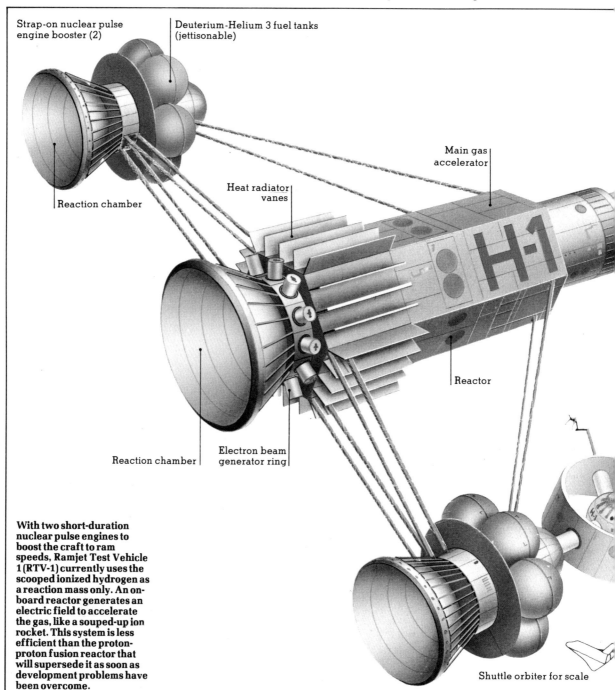

Strap-on nuclear pulse engine booster (2)

Deuterium-Helium 3 fuel tanks (jettisonable)

Main gas accelerator

Reaction chamber

Heat radiator vanes

Reactor

Reaction chamber

Electron beam generator ring

With two short-duration nuclear pulse engines to boost the craft to ram speeds, Ramjet Test Vehicle 1 (RTV-1) currently uses the scooped ionized hydrogen as a reaction mass only. An on-board reactor generates an electric field to accelerate the gas, like a souped-up ion rocket. This system is less efficient than the proton-proton fusion reactor that will supersede it as soon as development problems have been overcome.

Shuttle orbiter for scale

around the reactor inlet funnel ionized hydrogen by creating an enormous magnetic cone with an intake 100 km in diameter.

The ramjet only becomes effective at quite high speeds, so that an initial propulsion system is necessary. Present test vehicles use strap-on nuclear pulse engines adapted from the Daedalus-series probes (see p. 9 4). At 2 per cent of the speed of light the ramjet achieves 50 per cent efficiency, although the present experimental ramscoop has not yet achieved this.

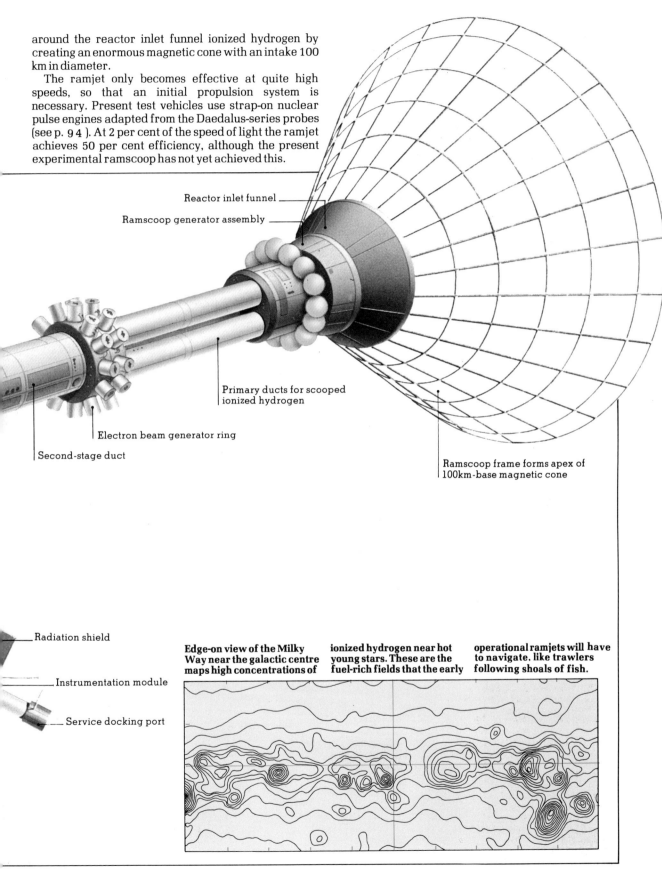

Reactor inlet funnel

Ramscoop generator assembly

Primary ducts for scooped ionized hydrogen

Electron beam generator ring

Second-stage duct

Ramscoop frame forms apex of 100km-base magnetic cone

Radiation shield

Instrumentation module

Service docking port

Edge-on view of the Milky Way near the galactic centre maps high concentrations of ionized hydrogen near hot young stars. These are the fuel-rich fields that the early operational ramjets will have to navigate, like trawlers following shoals of fish.

Lunar module The first small step

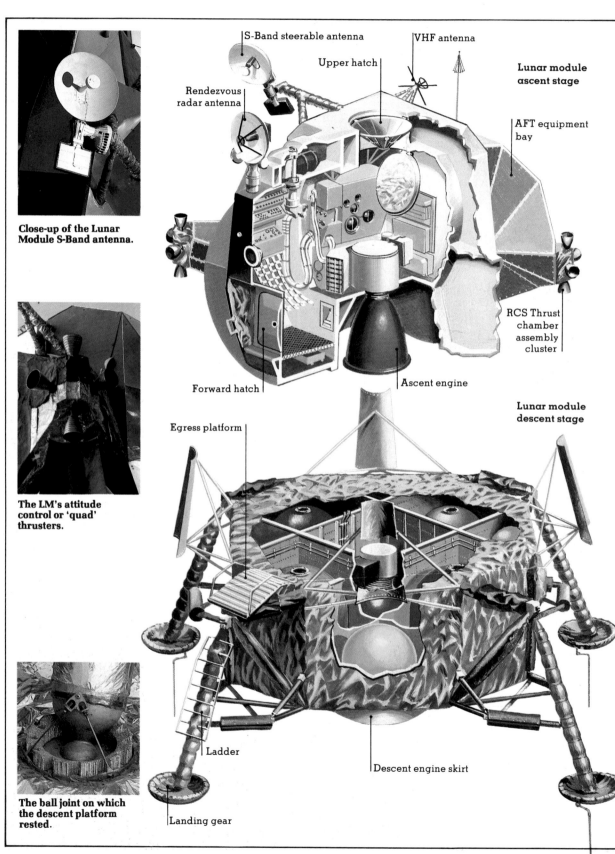

Close-up of the Lunar Module S-Band antenna.

The LM's attitude control or 'quad' thrusters.

The ball joint on which the descent platform rested.

S-Band steerable antenna

VHF antenna

Upper hatch

Rendezvous radar antenna

Lunar module ascent stage

AFT equipment bay

RCS Thrust chamber assembly cluster

Forward hatch

Ascent engine

Lunar module descent stage

Egress platform

Ladder

Descent engine skirt

Landing gear

The Lunar Module (left), was capable of operating as a separate craft for forty-eight hours which allowed thirty-five hours of activity on the Moon. In terms of systems and equipment the ascent stage was similar to the command module. It was almost 4 meters high. The descent stage was just over 3 meters high with legs extended and.
Right: the Apollo 11 LM *Eagle* leaves the Moon as Earth rises above the horizon.

THE GRUMMAN LUNAR MODULE was the first manned vehicle to land on the Moon in the Sea of Tranquility on the Apollo 11 mission. A completely self-sufficient spacecraft, equipped with all the subsystems necessary for life-support, control, guidance, navigation and communications, it consisted of two stages: the descent (lower) stage, with a gimballed, throttleable rocket engine, and an ascent (upper) stage, with its own rocket engine and the crew compartment.

Coupled to the Command Module in a turn-around docking manoeuvre the Lunar Module was then entered by two of the three Apollo astronauts. They piloted the vehicle to a pre-selected site, landed, completed their ground activities, and returned in the ascent stage, using the descent stage as a launching platform.

The ascent engine, which burnt a weight of fuel — just over 2265 kg equivalent to the weight of the ascent stage.

A joint on one of the legs of the LM's descent stage (above). Right: controls and displays for the main engine and for the attitude thrusters of the Lunar Module were duplicated. In this way either crewman in the two-man cabin could pilot the vehicle.

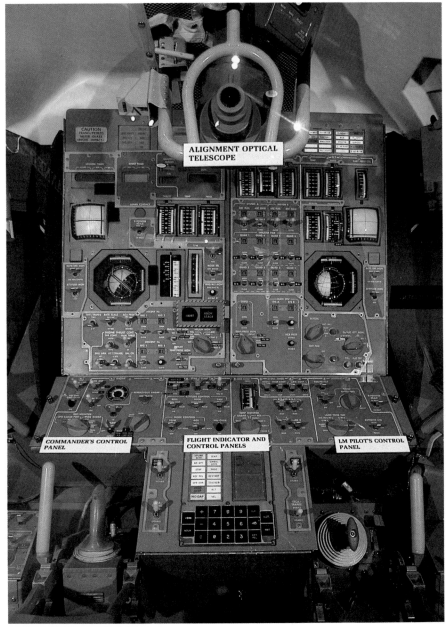

Lunar landers A regular people's service

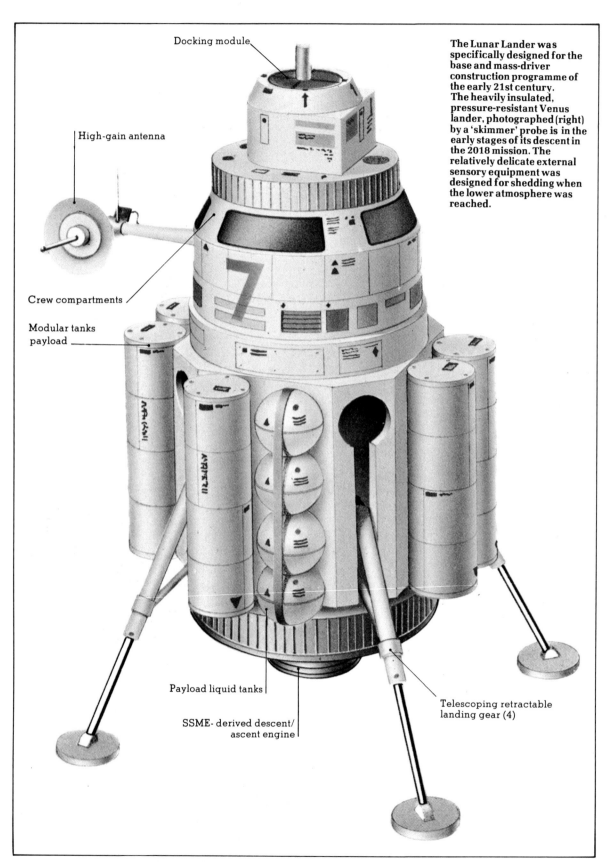

Docking module

High-gain antenna

Crew compartments

Modular tanks payload

Payload liquid tanks

SSME- derived descent/ ascent engine

Telescoping retractable landing gear (4)

The Lunar Lander was specifically designed for the base and mass-driver construction programme of the early 21st century. The heavily insulated, pressure-resistant Venus lander, photographed (right) by a 'skimmer' probe is in the early stages of its descent in the 2018 mission. The relatively delicate external sensory equipment was designed for shedding when the lower atmosphere was reached.

Space colonies Pioneering enthusiasm

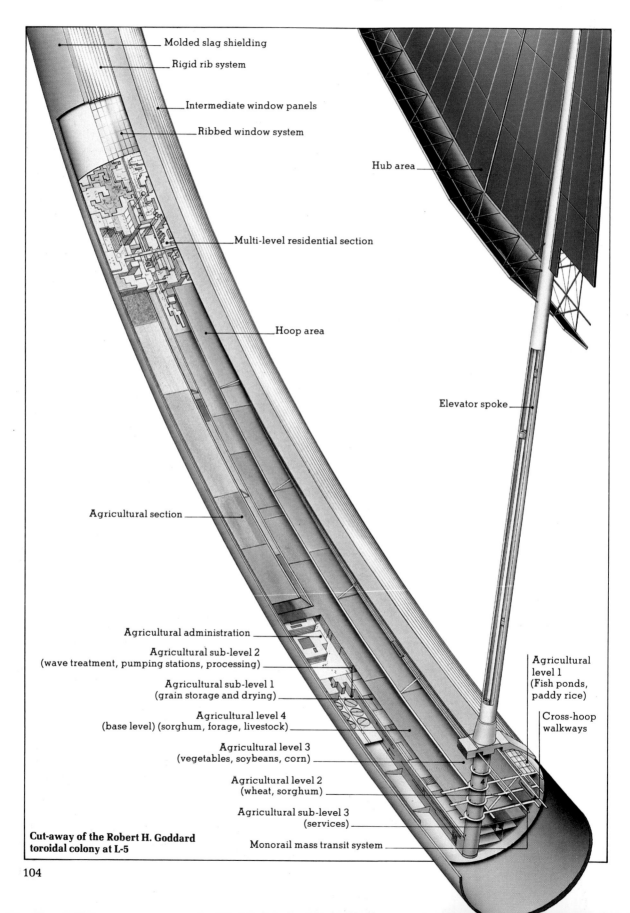

Molded slag shielding

Rigid rib system

Intermediate window panels

Ribbed window system

Hub area

Multi-level residential section

Hoop area

Elevator spoke

Agricultural section

Agricultural administration

Agricultural sub-level 2
(wave treatment, pumping stations, processing)

Agricultural sub-level 1
(grain storage and drying)

Agricultural level 4
(base level) (sorghum, forage, livestock)

Agricultural level 3
(vegetables, soybeans, corn)

Agricultural level 2
(wheat, sorghum)

Agricultural sub-level 3
(services)

Monorail mass transit system

Agricultural
level 1
(Fish ponds,
paddy rice)

Cross-hoop
walkways

**Cut-away of the Robert H. Goddard
toroidal colony at L-5**

THE HISTORY OF THE COLONIES is unusual in that they were proposed and largely designed *before* there was any real demand for their existence. That they were built when they were, as early as the 2030s, is a tribute to the persistence and enthusiasm of the pressure· groups and individual proponents who gathered under NASA's umbrella in the last two decades of the 20th century. At the time, the lack of finance for the American space programme encouraged NASA to look for ways of harvesting public support for the future of space. The idea of the Colonies was thought to be ideal, as it would clearly involve many members of the general public in a very personal way. This grass-roots development of space may have had slim economic justification, as has frequently been argued, but there seems little doubt that it has helped popularize space travel and make other space projects palatable to the public. Certainly, the Budgetary Office of the International Space Administration considers the Colonies to be politically important and has traditionally been generous with them.

Discounting the workforce that would in any case be necessary to maintain the various energy and communications programmes, the economic viability of the Colonies is still only marginal, and there is no certainty that this will ever improve. Basically, the Colonies survive on the will and enthusiasm of their inhabitants, and a strong sense of pioneering commitment is obvious to the visitor. Not that that is any less reason for their existence, but from the very start, the Colonies have been justified by their respective nations on cost-effective grounds, somewhat defensively. It is as if decades of hard bargaining for governmental budgets has left the Colonies unwilling to admit that the main motivation is to create experimental societies in space.

All the present Colonies are variants of the O'Neill Torus, resembling a wheel, with the majority of the population living in the rim under simulated gravity created by spin (popularly called 'G-sim'). In fact, O'Neill, who first championed the cause of space colonies with any real success, had envisaged a hierarchy of constructions, developing as technological ability progressed and as more living room was needed. Taking into account mass-ratio and the amount of shielding that would be necessary against solar flares, the ultimate large colony would be a cylinder. The intermediate stage, one that could be constructed with a lower-grade technology, would be the so-called Bernal Sphere. The first construction of all would be torus-shaped (like a doughnut).

To date, the O'Neill Torus is the only functioning design, although the ISA is currently in the middle of a long-term construction project to complete the first Bernal Sphere (funding difficulties have caused the ISA to delay the final stages and, for the time being at least, the existing structure is in use purely as a hydroponics farm).

There are, at the moment, five toroidal colonies — two American (the Robert H. Goddard and the Richard M. Nixon), one Russian (the Konstantin Tsiolkovsky), one Brazilian (the Pedro Carvejal), and one pan-European (still provisionally known as Europa I, due to the customary inability of the European Parliament to reach an agreement). All except the Russian Colony were assembled with ISA facilities and guidance.

Visiting restrictions are minimal, in accordance with Article XIII of the 1967 Treaty, 2222/XX, although hotel accommodation is limited (reservations are automatically made by Shuttle lines that fly Lagrange routes). Unfortunately, in apparent contravention of Article XIII, the Russian Colony has been declared a part of the national territory, this being justified on the grounds that it contains installations of military importance (itself a violation of Article IV of the 1967 Treaty). Because of this, normal visas are required to visit Konstantin Tsiolkovsky, and these are not always readily granted.

Allocation of space in the colony

Residential

Transportation

Offices

Shops

Hospital

Storage

Schools

Assembly

Indoor recreation

Service industry community functions

Public open space

Projected area, m² 43 x 10⁴

Agriculture & processing

Mechanical & life support subsystems

20 x 10⁴   4 x 10⁴

Ground vehicles The Lunar Rover

FIRST OF ALL extra-terrestrial ground vehicles, the Apollo programme's Lunar Roving Vehicle set the design style for later models. A two-man, electrically powered, four-wheel drive car, the Rover made its debut in 1971 in the Hadley Rille area of the Moon.

Its power sources were two 36-volt silver-zinc batteries, and its operational lifetime 78 hours during the lunar day, giving it the ability to cover up to 92 kms. So efficient was the Lunar Rover that it could carry 400kg Earth-weight — more than twice its own weight (for comparison, an average family vehicle can carry half its weight). Fully loaded, it could climb a 25° slope, and on level ground could reach 16 km/hr.

Interestingly, because of the absence of a magnetic field on the Moon, an original navigational system had to be worked out by Boeing, the contractors. Essentially, the course was set initially by map, and then maintained by a mixture of dead-reckoning (odometer on the wheels and elapsed-time measurements), a directional gyro compass (which because of its spin retained the same orientation regardless of the Rover's turns), and a sophisticated sundial.

Designed at a time when Earth-bound cars were built for looks and speed, the Lunar Rover, the most dramatic car of its time, excelled in neither of these departments.
Below left: the stages of deployment of the Lunar Rover.

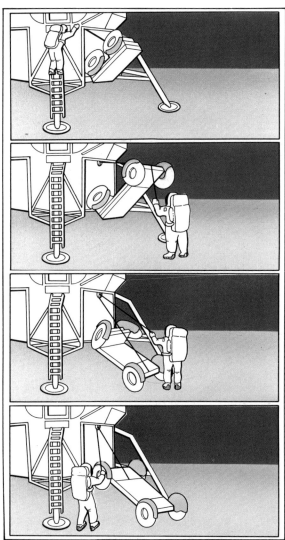

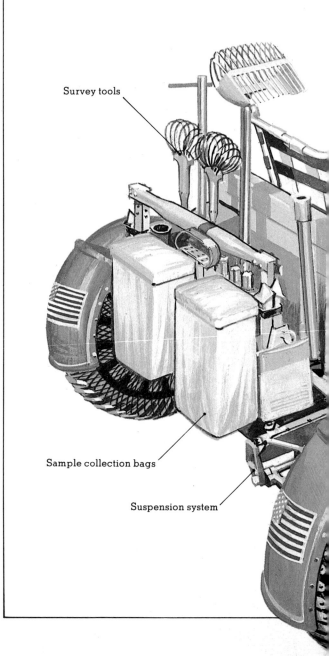

Survey tools

Sample collection bags

Suspension system

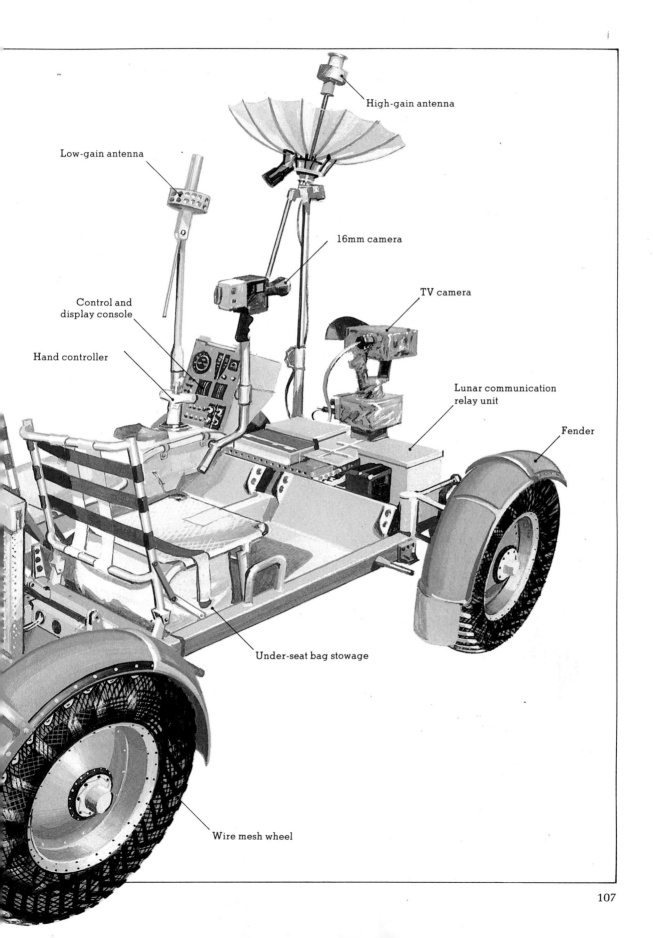

High-gain antenna

Low-gain antenna

16mm camera

TV camera

Control and
display console

Hand controller

Lunar communication
relay unit

Fender

Under-seat bag stowage

Wire mesh wheel

Ground vehicles The Lunar Rover

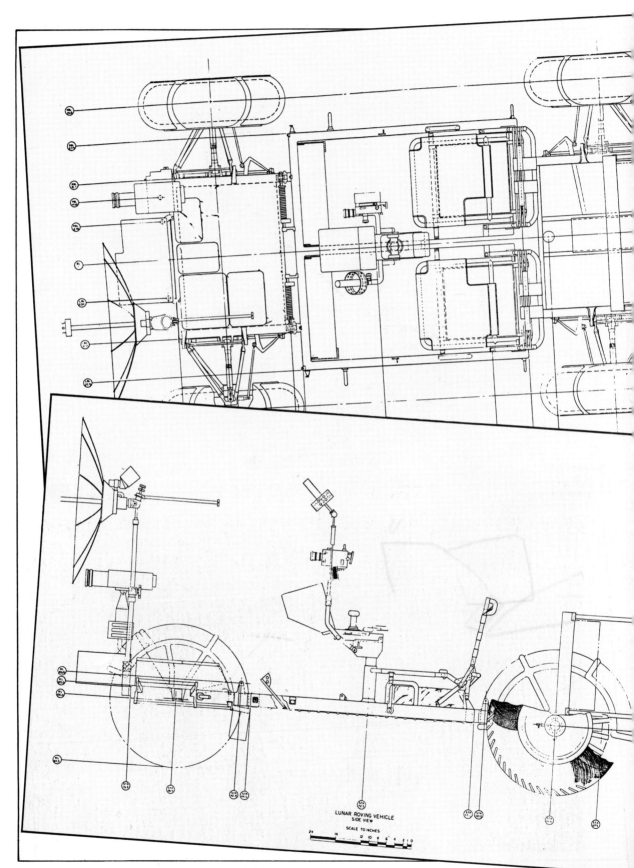

LUNAR ROVING VEHICLE
SIDE VIEW

SCALE TO INCHES

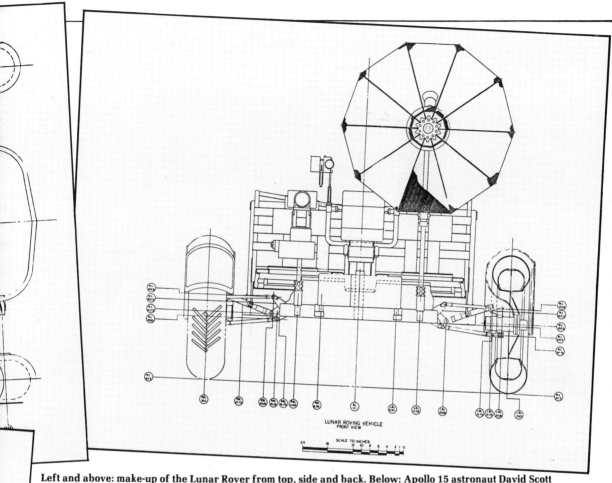

LUNAR ROVING VEHICLE
FRONT VIEW

SCALE TO INCHES

Left and above: make-up of the Lunar Rover from top, side and back. **Below:** Apollo 15 astronaut David Scott keeps the engine ticking over for James Irwin, to take him back to the module 'Falcon' after collection of some of their 82 kg of samples.

Advanced Rovers Customized for a planet

THE DESIGN OF SURFACE VEHICLES for planetary use has now become an engineering discipline in its own right, and a number of firms specialize in producing customized rovers. As demand is still relatively limited, there is no question of mass-production except for mesh tyres, track units, fuel cells and certain minor parts; vehicles are invariably purpose-built. However, the new ISA contract for 50 lunar trailer trucks awarded to General Motors may herald the start of assembly-line production.

Within the Inner System, solar power is viable, and is normally used when night-time operation is not essential; on Mars, solar power is only marginally efficient for vehicles. Beyond Mars-orbit, all ground vehicles must use self-contained power sources, meaning fuel cells. For high-powered transport over long distances, sub-orbital rockets are the only practical answer.

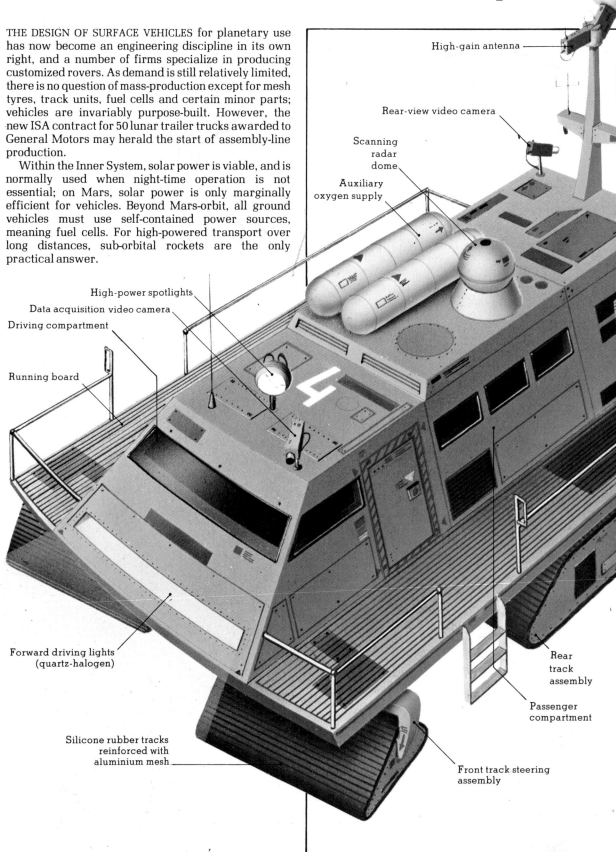

High-gain antenna

Rear-view video camera

Scanning radar dome

Auxiliary oxygen supply

High-power spotlights

Data acquisition video camera

Driving compartment

Running board

Forward driving lights (quartz-halogen)

Silicone rubber tracks reinforced with aluminium mesh

Rear track assembly

Passenger compartment

Front track steering assembly

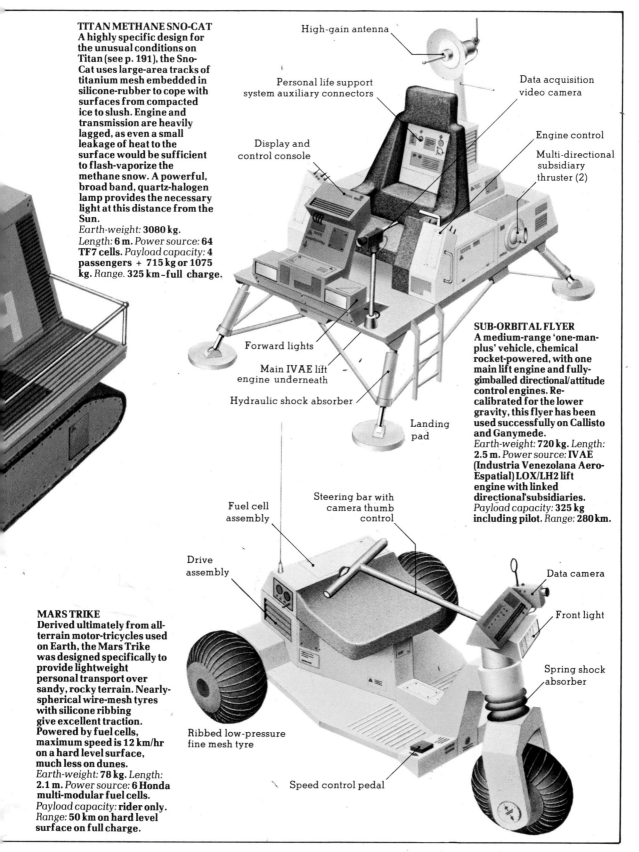

TITAN METHANE SNO-CAT
A highly specific design for
the unusual conditions on
Titan (see p. 191), the Sno-
Cat uses large-area tracks of
titanium mesh embedded in
silicone-rubber to cope with
surfaces from compacted
ice to slush. Engine and
transmission are heavily
lagged, as even a small
leakage of heat to the
surface would be sufficient
to flash-vaporize the
methane snow. A powerful,
broad band, quartz-halogen
lamp provides the necessary
light at this distance from the
Sun.
Earth-weight: **3080 kg.**
Length: **6 m.** *Power source:* **64
TF7 cells.** *Payload capacity:* **4
passengers + 715 kg or 1075
kg.** *Range.* **325 km–full charge.**

High-gain antenna

Personal life support
system auxiliary connectors

Data acquisition
video camera

Display and
control console

Engine control

Multi-directional
subsidiary
thruster (2)

Forward lights

Main IVAE lift
engine underneath

Hydraulic shock absorber

Landing
pad

SUB-ORBITAL FLYER
A medium-range 'one-man-
plus' vehicle, chemical
rocket-powered, with one
main lift engine and fully-
gimballed directional/attitude
control engines. Re-
calibrated for the lower
gravity, this flyer has been
used successfully on Callisto
and Ganymede.
Earth-weight: **720 kg.** *Length:*
2.5 m. *Power source:* **IVAE
(Industria Venezolana Aero-
Espatial) LOX/LH2 lift
engine with linked
directional subsidiaries.**
Payload capacity: **325 kg
including pilot.** *Range:* **280 km.**

Fuel cell
assembly

Steering bar with
camera thumb
control

Drive
assembly

Data camera

Front light

Spring shock
absorber

MARS TRIKE
Derived ultimately from all-
terrain motor-tricycles used
on Earth, the Mars Trike
was designed specifically to
provide lightweight
personal transport over
sandy, rocky terrain. Nearly-
spherical wire-mesh tyres
with silicone ribbing
give excellent traction.
Powered by fuel cells,
maximum speed is 12 km/hr
on a hard level surface,
much less on dunes.
Earth-weight: **78 kg.** *Length:*
2.1 m. *Power source:* **6 Honda
multi-modular fuel cells.**
Payload capacity: **rider only.**
Range: **50 km on hard level
surface on full charge.**

Ribbed low-pressure
fine mesh tyre

Speed control pedal

LUNAR SUB-ORBITAL BUS
This long-range vehicle for up to 12 passengers operates on low trajectories, using chemical rockets. Fore and aft lift engines deliver vertical thrust, whilst two gimbal-mounted engines at the rear control directional flight and two braking engines underneath the cockpit permit controlled descent. This seemingly extravagant array of engines gives exceptional manoeuvrability, at the same time allowing the pilot to maintain a constant attitude. Skids on universal joints facilitate landing on a wide variety of terrain.
Earth-weight: **3 420 kg.**
Length: **13.75 m.** *Power source:* **4 Rolls-Royce LA27 lift engines, 4 Rolls-Royce LAH19 directional engines.** *Payload capacity:* **12 passengers + 650 kg or 1370 kg.** *Range:* **No limit to single-leg journeys, but engine-burn time is limited to 120 secs on lift engines/190 secs on directional engines (fuel tanks are shared).**

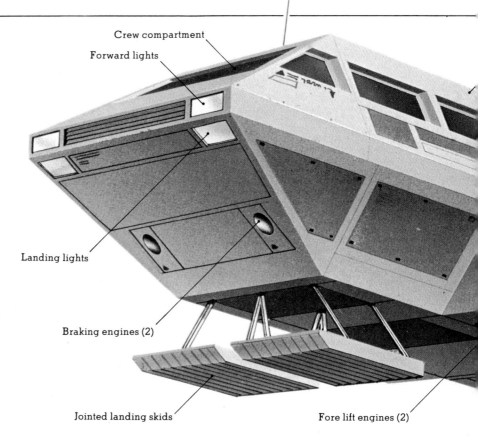

Crew compartment

Forward lights

Landing lights

Braking engines (2)

Jointed landing skids

Fore lift engines (2)

LUNAR TRACKED BUS
First introduced at Clavius Base in 2003, the tracked bus proved to be a popular vehicle, providing a shirt-sleeve environment for passengers in transit between Base buildings. The use of tracks instead of the more normal mesh tyres was made possible by recent improvements in fuel cell efficiency (see p. 114); the only limitation is the rather low speed — 10 km/hr.
Earth-weight: **2,735 kg.**
Length: **6.3 m.** *Wheel track:* **2.1 m.** *Power source:* **2 banks of 12 LR Industries.** *Payload capacity:* **6 passengers + 500 kg or 860 kg fuel cells.** *Range:* **180 km on full charge.**

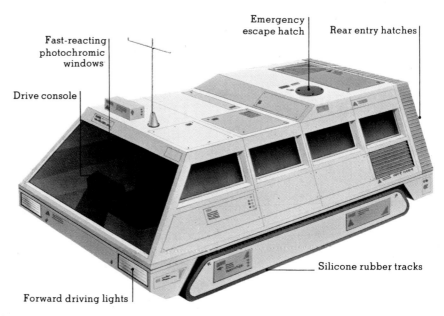

Fast-reacting photochromic windows

Emergency escape hatch

Rear entry hatches

Drive console

Silicone rubber tracks

Forward driving lights

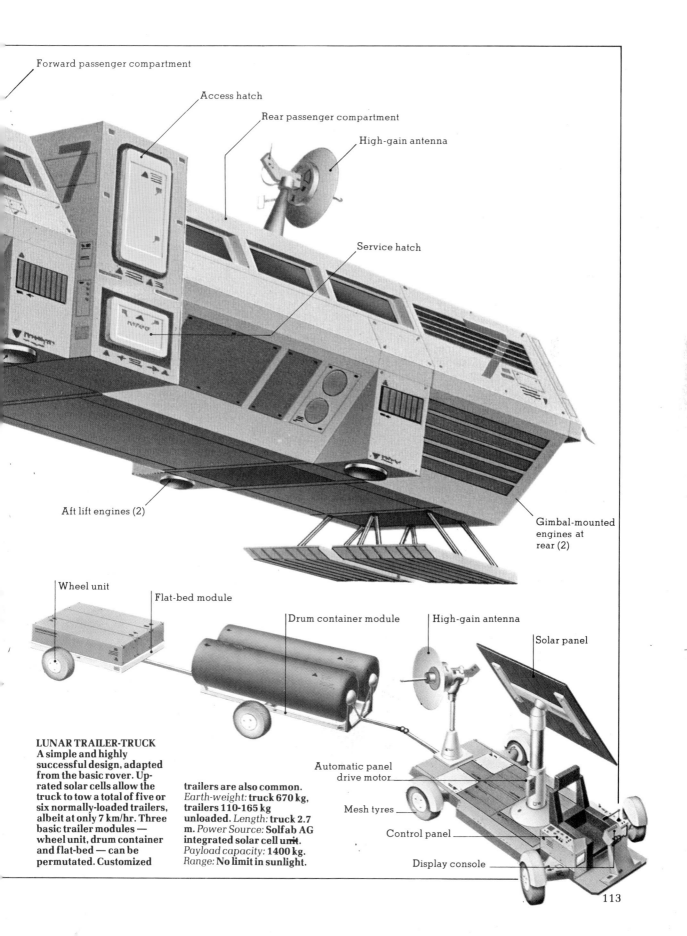

Forward passenger compartment

Access hatch

Rear passenger compartment

High-gain antenna

Service hatch

Aft lift engines (2)

Gimbal-mounted engines at rear (2)

Wheel unit

Flat-bed module

Drum container module

High-gain antenna

Solar panel

Automatic panel drive motor

Mesh tyres

Control panel

Display console

LUNAR TRAILER-TRUCK
A simple and highly successful design, adapted from the basic rover. Up-rated solar cells allow the truck to tow a total of five or six normally-loaded trailers, albeit at only 7 km/hr. Three basic trailer modules — wheel unit, drum container and flat-bed — can be permutated. Customized trailers are also common. *Earth-weight:* **truck 670 kg, trailers 110-165 kg** unloaded. *Length:* **truck 2.7 m.** *Power Source:* **Solfab AG integrated solar cell unit.** *Payload capacity:* **1400 kg.** *Range:* **No limit in sunlight.**

Power systems On-board energy

STILL STANDARD on many spacecraft, fuel cells are the basic method of generating electrical energy for general use on board. Whilst more sophisticated systems are available, such as tapping surplus nuclear energy on board interplanetary nuclear rockets, the beauty of the fuel cell lies in its simplicity and independence.

Baically, a fuel cell is a battery converting the energy released in a chemical reaction directly to electrical energy. In this example, the Apollo cell, the chemical reaction in question is the combination of hydrogen and oxygen, incidentally forming water but, more important, releasing energy. Energy is released because electrons in a water molecule are in a lower energy state than electrons in a gas molecule. From 50 to 60 per cent of this surplus energy is converted directly into electrical energy, and a welcome by-product is drinking water.

This fuel cell powerplant contains 31 individual fuel cells and operates at 27-31 volts, with a maximum output of 2300 watts. To save space, it is normal to store the reactants (H and O_2) as liquids, which means keeping the hydrogen at —253°C and the oxygen at —173°C.

Radioisotope Thermoelectric Generators (RTGs)

Normally used to power equipment that has small energy needs, RTGs have found a special place in unmanned spacecraft and remote equipment. The SNAP-27 shown here was used to power Apollo lunar scientific experiments, and the much larger RTG powered the systems on board the Out-Planet probe,

The fuel cells shown here depend, in the one case, on an internal but very long-lasting source of energy, and, in the other, on an external but all-pervading source, in the Solar System at least. Both these methods of propulsion have the advantage that, while the power they generate is not initially considerable, they are extremely reliable over long times and distances. The Apollo fuel cell (above) was a British invention. It powered the spacecraft from just before lift-off till separation of the service-module just before splashdown. The cells had to be 'purged' with liquid hydrogen from time to time to clean out impurities. They also charged the emergency batteries. Left: an RTG in use during an Apollo 14 experiment. Right: a section of a solar panel as used on Skylab (insert).

Voyager, on its long journey to the limits of the Solar System.

Basically, an RTG generates electrical power by the decay of a radioactive isotope. The fuel in SNAP-27 was plutonium 238, in a 3.8kg capsule that heats up to 600°C on the inside due to the radioactive decay. The thermoelectric element, which is a loop made of two conductors whose energy levels change when the temperature changes, converts this heat energy into electrical energy. The initial output of SNAP-27 was only 70 watts, but is so reliable that it was still producing 90 per cent of this output ten years later. The vanes are for cooling.

The larger RTGs of Voyager generated 1800 watts of heat, also from plutonium; thermocouples converted this to 400 watts of electrical power.

Solar Arrays

The popularity of solar power as an energy source has increased as the technology has improved. Basically, silicon solar cells are banked together in arrays such as the panel used on the Skylab space station.

The solar cell works like this: each square of silicon is impregnated with phosphorus and boron impurities to create two electrically-different sections. These are like the positive and negative plates of a battery. Each cell produces only about 60 milliamps at 0.4 volts, so thousands are needed to power even a moderately sized spacecraft.

In the case of Skylab, for example, 147,840 cells were clustered on each 'wing', exposing 219 sq meters to the rays of the Sun, and capable of generating 10,500 watts at 55°C.

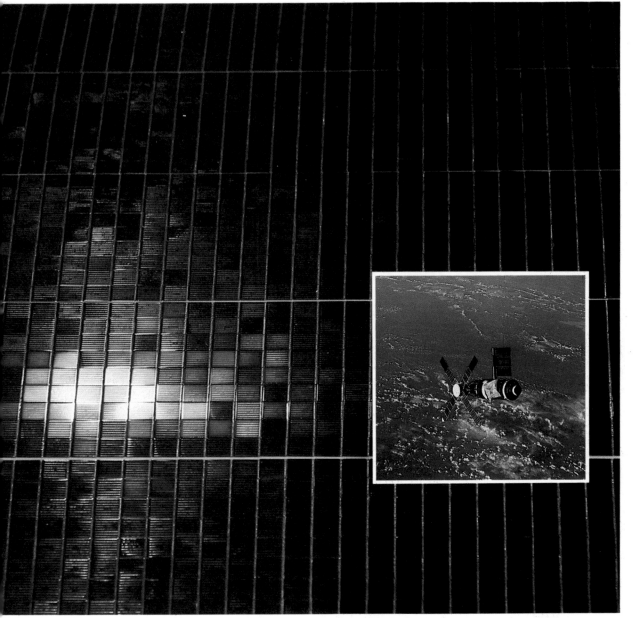

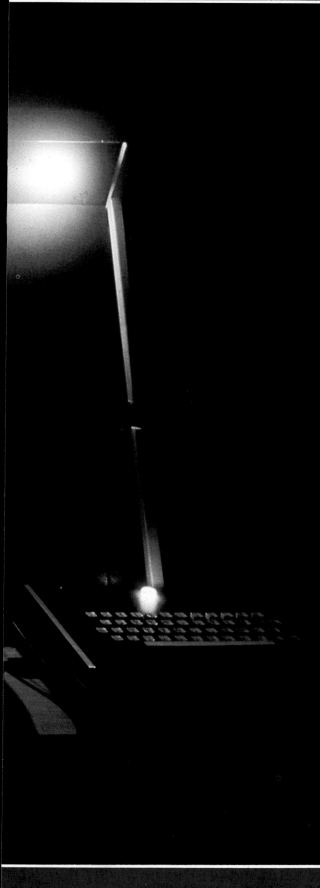

Control and guidance

THE ARCANE MYSTERIES of control and guidance appear to most people to be shrouded in complexities. Certainly, modern astronavigation bears little resemblance to sailing a dinghy up an estuary, and although comparisons are inevitably drawn between plotting the trajectory of a space vehicle and laying a course at sea, beginners are warned not to take the analogy too literally. Nevertheless, the laws that determine the paths of suns, planets and spacecraft are reasonably straight forward, and now that adaptable computer programmes have taken the tedium out of calculation, it is perfectly possible for the layman to acquire a working background knowledge without losing too much sleep.

The section that follows is in the nature of a short primer and, while it is in no way intended to qualify anyone to direct even a spacetug between orbits, it may at least enable the passenger of a modern spacecraft to understand the logic of the pilot's actions, and thus be able to take a more informed interest in the journey.

En route from Mars, the astronavigator of a large nuclear-thermal craft monitors the final approach to Earth on his console.

Orbital mechanics 3-D navigation

IT IS TEMPTING to extend the analogy between space and the sea, especially for the navigator. His problem is essentially the same in both cases — laying a course between points over sometimes large distances, allowing not only for the performance of his ship but also for the various external forces.

However, there are two major differences between sea and space navigation. First there is the scale — interplanetary journeys bear no comparison with even Columbus' crossing of the Atlantic. The other difference is the effect of gravity — every body in space, from whole galaxies down to the smallest particle, exerts a gravitational pull on neighbouring bodies. The strength of this pull depends on the mass of the body and on its distance. A helpful analogy is to think of space as a flexible membrane stretched flat. The Sun and planets each make depressions due to their mass, creating a series of gravity 'wells'. Moving about in space involves crossing the slopes of these 'wells'.

The net result of all these gravitational fields is that every body in space, from planets to spacecraft, obeys the laws of motion and follows an orbit. For spacecraft, which have their own propulsion system, it is customary to call this orbit a trajectory. There are just four kinds of orbit — circular, elliptical, parabolic and hyperbolic (actually, a circle is a special form of an ellipse). Together, these are called conic orbits, because they can all be constructed by slicing through a cone.

A spacecraft or satellite orbiting the Earth with just enough speed to stay up will be in a circular orbit. The pull of the Earth's gravity and the spacecraft's velocity are just balanced. Now, if the main thrust motors are fired, the ship's velocity is increased, and the orbit becomes elliptical. By giving another burn at the orbit's closest point to the Earth — called the perigee — the ellipse can be made more eccentric. With an even stronger push, the orbit becomes a parabola, which is no longer a closed orbit. If the 'perigee kick', as it is called, is stronger still, the orbit becomes a hyperbola, the fastest of all orbits (see p.120).

The planets, of course, are all in closed orbits around the Sun, which means that they follow elliptical paths. Most are fairly circular although some, like Mercury's and Pluto's, are quite eccentric. Fortunately, all the planetary orbits, except Pluto's, lie on roughly the same plane — calculations involving inclined orbits pose an additional complication.

The astronavigator's computational problems really start when he has to allow for the influence of neighbouring bodies. A spacecraft orbiting one isolated planet would be a simple case, a two-body problem. In reality, though, there are other things to take into account. A spacecraft in lunar orbit, for example, obviously interacts with the Moon. But both are strongly affected by the Earth, and all three are under the even stronger influence of the Sun. Apart from this, the actual orbit of the Earth-Moon system is influenced by the other planets in the Solar System. There are no simple solutions to all this, and in fact only a two-body problem has what is called an unrestricted solution. Special cases of the three-body problem have been solved, but for the most part, the calculations for each situation must be worked out from scratch.

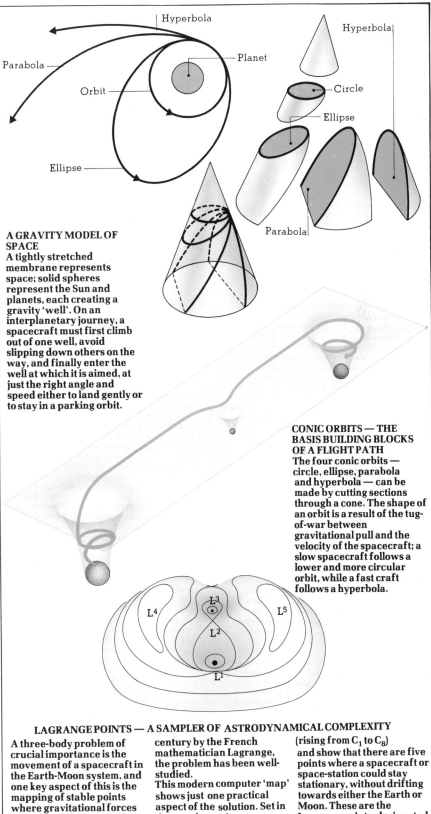

A GRAVITY MODEL OF SPACE
A tightly stretched membrane represents space; solid spheres represent the Sun and planets, each creating a gravity 'well'. On an interplanetary journey, a spacecraft must first climb out of one well, avoid slipping down others on the way, and finally enter the well at which it is aimed, at just the right angle and speed either to land gently or to stay in a parking orbit.

CONIC ORBITS — THE BASIS BUILDING BLOCKS OF A FLIGHT PATH
The four conic orbits — circle, ellipse, parabola and hyperbola — can be made by cutting sections through a cone. The shape of an orbit is a result of the tug-of-war between gravitational pull and the velocity of the spacecraft; a slow spacecraft follows a lower and more circular orbit, while a fast craft follows a hyperbola.

LAGRANGE POINTS — A SAMPLER OF ASTRODYNAMICAL COMPLEXITY

A three-body problem of crucial importance is the movement of a spacecraft in the Earth-Moon system, and one key aspect of this is the mapping of stable points where gravitational forces are balanced. First examined in the 18th century by the French mathematician Lagrange, the problem has been well-studied.

This modern computer 'map' shows just one practical aspect of the solution. Set in the x, y plane, the contours show potential energy (rising from C_1 to C_8) and show that there are five points where a spacecraft or space-station could stay stationary, without drifting towards either the Earth or Moon. These are the Lagrange points, designated L_1 to L_5.

Flight profiles Orbit to orbit

THE REAL WORK starts with the filing of the flight profile. There are standard procedures for doing this agreed between the various space authorities signatory to the 2028 Convention. It is normal practice first to state the mission. This will determine the choice of trajectories, which will have to take account of fuel/velocity considerations and the launch window. Almost inevitably, an interplanetary journey involves going from one orbiting body to another, be they planets, asteroids or space-stations. The problem, therefore, is to transfer the spacecraft from one orbit to another.

Transfer Orbits

These are the basic elements of the flight profile. As an example, the most popular interplanetary route is the Earth-Mars transfer. Whatever form the launch takes, the spacecraft must accelerate to a faster orbit than the Earth's, one that will intersect with the orbit of Mars. The correct timing will ensure that the intersection takes place when Mars is actually there! The cheapest transfer orbit, using least fuel and energy, is an ellipse which just grazes both the orbits of both Earth and Mars. This is the well-known Hohmann Transfer Orbit, much used by commercial craft. Of course, as the table shows, it makes for a very long journey. Faster transfers are possible, in parabolic and hyperbolic orbits, but the astronaut's constant preoccupation with fuel supplies means that these fast orbits are rarely used in the normal course of interplanetary travel.

Calculating Trajectories

Naturally, the on-board computer will make all the necessary calculations, and these are fed directly to the ship's control and guidance systems. However, there exist various rule-of-thumb methods for working out trajectories; these can give the astronavigator a broad picture of his mission. This is particularly useful in spacecraft equipped with primitive computers.

The basic rough-and-ready system used throughout the Solar System is the Patched Conic Method. It goes far towards removing the headache of dealing with several gravitational fields at once. The principle is straightforward — the trajectory is roughly divided into 'legs', each dealing with only one planetary body at a time. Every planet is assigned a 'sphere of influence', the area in which that planet is the principal attraction. The Earth's sphere of influence is approximately one million kilometers. Outside that sphere the Sun becomes the main attracting force.

So, using the Patched Conic Method to calculate an Earth-Mars trajectory, we would split the journey into three legs: from Earth to the edge of its sphere of influence, then the longest leg within the sphere of influence of the Sun, and finally the leg within Mars' sphere of influence.

Of course, such do-it-yourself trajectory planning is becoming rare these days, now that quite sophisticated tape programmes can be bought for the most common trajectories in the Solar System. These can be adapted for each journey.

HOHMANN TRANSFER ORBITS – ECONOMY CLASS

First proposed by Walter Hohmann in 1925, the Hohmann Transfer Orbit connects two orbits with the least expenditure of energy. The ellipse just touches both planetary orbits.

Hohmann journey from the inner to the outer Solar System

Target Planet	Flight time from Earth	Spacecraft velocity relative to Earth (km/sec)
Mercury	107 days	7.5
Venus	146 days	2.5
Mars	260 days	2.9
Jupiter	2.7 years	8.8
Saturn	6 years	10.3
Uranus	16 years	11.3
Neptune	30 years	11.6
Pluto	45 years	11.8

FAST TRANSFER ORBITS — AT A PRICE:
By increasing the launch velocity, the journey time can be reduced, but more fuel will be used up. To achieve the hyperbolic trajectory, the spacecraft would have to be accelerated well beyond the escape velocity of the Solar System, as this orbit is open.

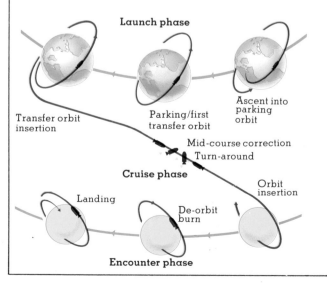

Launch phase

Transfer orbit insertion

Parking/first transfer orbit

Ascent into parking orbit

Mid-course correction
Turn-around

Cruise phase

Orbit insertion

Landing

De-orbit burn

Encounter phase

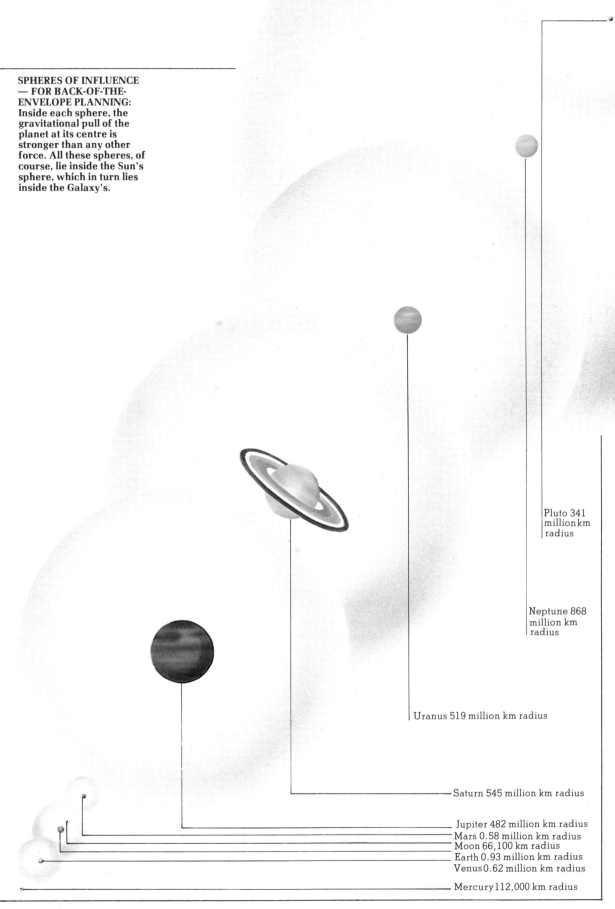

SPHERES OF INFLUENCE — FOR BACK-OF-THE-ENVELOPE PLANNING: Inside each sphere, the gravitational pull of the planet at its centre is stronger than any other force. All these spheres, of course, lie inside the Sun's sphere, which in turn lies inside the Galaxy's.

Pluto 341 million km radius

Neptune 868 million km radius

Uranus 519 million km radius

Saturn 545 million km radius

Jupiter 482 million km radius
Mars 0.58 million km radius
Moon 66,100 km radius
Earth 0.93 million km radius
Venus 0.62 million km radius

Mercury 112,000 km radius

Launch/Encounter The critical moments

WHEN THE LAUNCH and encounter phases are part of the same flight as the transfer, they are usually the critical moments, demanding the greatest concentration and care. Also, the launch itself may, on occasion, be used to put the spacecraft directly into the transfer orbit. Nowadays, however, as spacecraft designs become more and more specialized, it is more usual for the spacecraft to begin its journey from a parking orbit above a planet rather than directly from its surface.

Escape Velocity

Most planetary launches are now made by the various shuttle services that have developed to meet the growing traffic. The basic requirement, of course, is to exceed the planet's escape velocity, which for the Earth is 11 km/sec. If anything less than this is achieved the spacecraft will fall back to the surface. Naturally, large amounts of fuel are needed at launches. This is why rockets appear to leave the ground slowly at first, gradually picking up speed; what is happening is that, as the propellant is used up, the spacecraft becomes lighter and therefore accelerates at a faster rate.

Rocket engineers concentrate on two things to attain a high velocity. One is improving the 'mass-ratio', which is the ratio of a full rocket to an empty one; they do this by using lightweight materials and thin-walled fuel tanks. The other improvement is in the propellant. High-energy propellants are the most valued, although there seems little prospect of any major advance in chemical propellants. Kerosene-LOX, hydrogen-LOX and hydrogen-fluorine are all good choices, and most base filling-stations carry a choice of all three.

Multi-stage launches

Heavy-duty shuttles and rockets with small payloads to deliver can manage a launch from Earth in one stage, but the majority of ascents still use the multistage system. The advantage is that, when the first stage has burnt out, it can be jettisoned, and the remaining rocket has less weight to lift. The early lunar missions using three and even four stages are now considered quite extravagant, but two stages are common. The problem, for cost-conscious shuttle lines, is to recover the jettisoned stage in a re-usable condition.

Launch windows and corridors

The timing of the launch depends on factors such as the position of the target, the weather, port clearance, traffic, etc. Launch Base Administration is normally responsible for co-ordination. Corridors vary with the launching site and the local regulations.

Encounter phase

Ignoring, for the time being, a fly-by (which in manned spaceflight is really an orbital manoeuvre), the choices at the end of the journey are planetary capture and landing. In either case, the spacecraft will need to brake. This is done by turning the craft round as it nears the target planet, and then firing the main engine as a retro. This inserts the spacecraft into an orbit, usually a circular parking orbit. If landing is to follow, then the retro must be fired once more to reduce the velocity enough for the craft to fall towards the surface.

TYPICAL MULTI-STAGE LAUNCH:
The rocket uses fuel in order to gain height and thus becomes lighter. The fuel weight and the thrust at each point in the ascent must be determined, with a view to achieving the necessary acceleration, until the early, powerful stages can be allowed to fall away.

Spacecraft is launched into orbit almost horizontally. This is done by vanes that deflect exhaust gases or by mounting engine on gimbals

m_{p2}

$mm_{b2} + M_2$

Second stage

v_m

v

Jettisoned empty booster

m_b

m_p

First stage propels spacecraft almost vertically

$m_b + M$

Acceleration increases as propellant is used up

Slow acceleration to start with

Downrange splashdown

m_b mass of booster
M mass of propellant
m_p mass of payload
v velocity

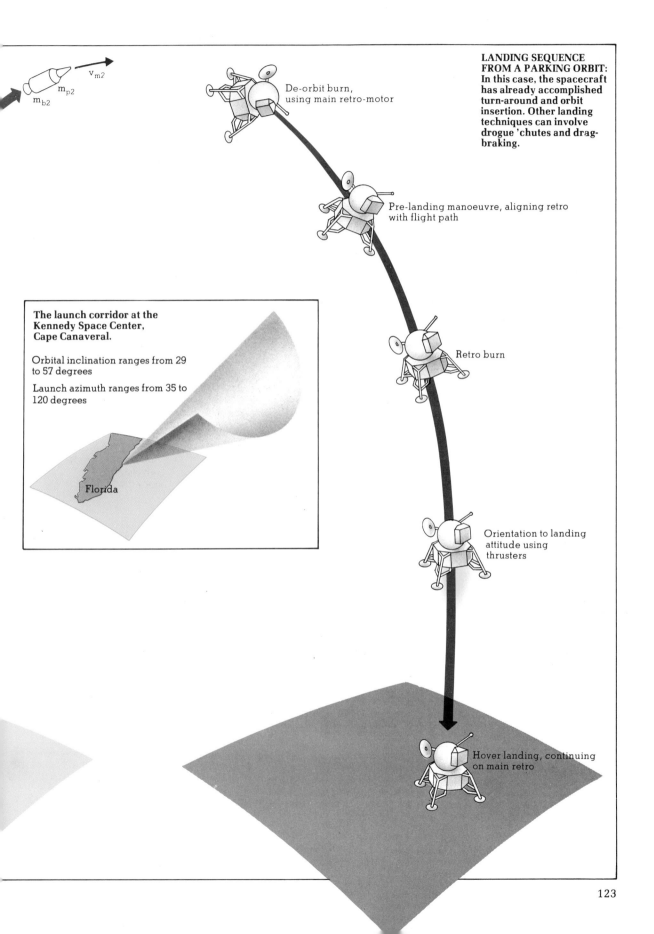

v_{m2}

m_{p2}

m_{b2}

De-orbit burn,
using main retro-motor

**LANDING SEQUENCE
FROM A PARKING ORBIT:**
In this case, the spacecraft
has already accomplished
turn-around and orbit
insertion. Other landing
techniques can involve
drogue 'chutes and drag-
braking.

Pre-landing manoeuvre, aligning retro
with flight path

**The launch corridor at the
Kennedy Space Center,
Cape Canaveral.**

Orbital inclination ranges from 29
to 57 degrees

Launch azimuth ranges from 35 to
120 degrees

Florida

Retro burn

Orientation to landing
attitude using
thrusters

Hover landing, continuing
on main retro

Orbital manoeuvres Fighting gravity's pull

ANY CHANGES TO THE SPACECRAFT'S trajectory are classed as manoeuvres; these can cover the whole range from orbit insertion and changing the plane of the orbit to minor corrections to keep the ship on its flight path.

Changing orbits
The most basic manoeuvre is to change the shape of the spacecraft's orbit. So, moving out of the original parking orbit, whether to rendezvous with another spacecraft 100 km higher or to embark on an eight and a half month voyage to Mars, the astronaut will need to fire the main motors for exactly the right amount of time and in exactly the right place. Basically, the spacecraft can be moved into a faster, more eccentric orbit by being given a short, sharp burn at its closest approach to Earth. This is known as a 'perigee kick': the larger the kick, the faster the orbit. In exactly the same way, the orbit can be 'circularized' by firing the motor at the other end of the ellipse — an 'apogee kick'.

Changing the plane of the orbit
Much more complicated is the manoeuvre necessary to move the spacecraft into an orbit at a different

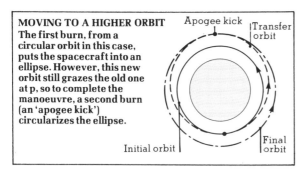

MOVING TO A HIGHER ORBIT
The first burn, from a circular orbit in this case, puts the spacecraft into an ellipse. However, this new orbit still grazes the old one at p, so to complete the manoeuvre, a second burn (an 'apogee kick') circularizes the ellipse.

Apogee kick | Transfer orbit

Initial orbit | Final orbit

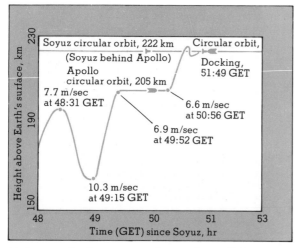

Early example of a spacecraft moving to a higher orbit was the Apollo-Soyuz rendezvous in 1975. In the final stages, Soyuz was in a circular orbit 222 km above the Earth, whilst Apollo was 17 km lower. The Apollo burn was timed so that it was slightly ahead of Soyuz; orbital speed at the higher altitude would be slower, so that, when Apollo was boosted above Soyuz, it fell gently back to the Russian craft.

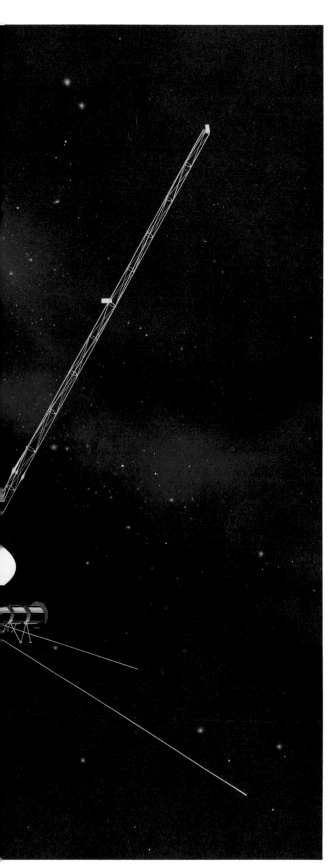

angle. The engine must be fired at exactly the point where the two orbits intersect; nothing else will do. The length of the burn is also critical.

Gravity Assists — the sling-shot technique

Without a doubt, the Gravity Assist is the most elegant manoeuvre in the book, and never fails to delight passengers. When performed with one of the major planets, the results can be phenomenal. The general idea is to allow the spacecraft to be drawn into the gravitational field of a planet, which itself will be moving faster than the spacecraft. The planet pulls the craft along with it, accelerating it and, if the encounter has been properly executed, whipping the spacecraft on its way as if from a catapult. At its best, with a really massive planet like Jupiter, this technique can turn a very average elliptical trajectory into a very enviable hyperbola. Of course, slowing the spacecraft down at its destination can often be a big problem, and on at least one occasion (the unfortunate case of the Ganymede Star in 2046) a miscalculation has resulted in a spacecraft being accelerated out of the Solar System on a fatal journey to the stars. Recent regulations on the filing of flight profiles are aimed at preventing this kind of accident.

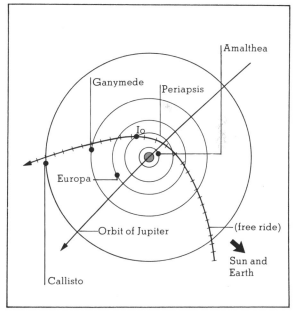

A FREE RIDE FROM A GIANT PLANET:
Using a Gravity Assist from Jupiter, the unmanned probe Voyager 1 (left) is whipped along and outwards to Saturn after its encounter in 1979. Getting fresh acceleration for nothing seems too good to be true. In point of fact, this acceleration is not strictly for nothing. The related loss is in the speed of Jupiter. It is, theoretically, slowed down by an amount too small to be measured.

Attitude control Roll pitch and yaw

THE BASIC REQUIREMENT for controlling a spacecraft, even before fixing its position along its trajectory, is to have it pointing in a known, stable direction. Any number of things can alter its attitude — the pressure of sunlight, for example, or even the movement of astronauts inside. For this reason, the spacecraft needs a device to sense precisely the way it 'sits'.

By convention, all spacecraft have three axes — x, y and z. Movements about these axes are called roll, pitch and yaw respectively. The two most common systems for controlling them are gyroscopes and star-tracking sensors. Often they are found together in the same ship, one backing up the other.

Gyroscopes
These sophisticated versions of a child's toy are used in threes, each one on a different axis, and mounted on gimbals. Friction inevitably causes them to 'drift', so that normally they must be checked by astronaut sightings of the Sun and stars. The gyroscopes are linked via the computer to attitude control thrusters —small jets in various positions on the spacecraft that can make small adjustments.

In some larger spacecraft, the gyroscope system has been developed even further, to make attitude adjustments itself, without the use of the thrusters. This method, called the 'control moment gyro' (CMG) system, was first used in Skylab in the 1970s. It employs three very large gyros, each with a motor-driven rotor; when the rotor is spun the gyro reacts to produce torque. This automatically compensates for small movements of the spacecraft.

Star-tracking sensors
Sailing enthusiasts will be keen to try out this method of navigation. The principles are the same as in maritime navigation, except, of course, that the astronaut has three dimensions to deal with. Nevertheless, the main task of the star-tracking equipment is to take a fix on two or more constant reference points. In practice, astronauts almost invariably use the Sun and Canopus. Canopus is convenient because it is not only the second brightest star in the sky, and therefore easy to identify, but it also lies at a very large angle from Sun — within 15° of ecliptic south (the ecliptic is the imaginary plane on which the Earth's orbit lies — a piece of interplanetary chauvinism that is very practical for astronauts!). Polaris is also sometimes used.

While sightings are sometimes made by astronauts, nowadays practically all star-tracking is done automatically. Even so, enthusiastic amateurs can experiment with a hand-held sextant and compare their efforts with the ship's computer. (WARNING: having adapted a sextant to space use, *never* point it directly at the Sun without using strong filters.)

The three axes of a spacecraft can describe any movement it makes. When the gyroscopes or star-tracker detect any untoward movement, they inform the computer, which calculates the amount of opposite thrust necessary to correct it. Two or more of the thrusters are then directed to burn, in the opposite direction to the movement.

THE QUAD JET

Thrusters come in all shapes and sizes, but what they all have in common is the ability to apply controlled movement around any axis. One of the most flexible systems is the quad jet. One unit gives thrust in four directions. If four of these units are placed strategically around the circumference of a spacecraft then they can take care of any desired movement.

The Apollo flights opted for the quad jet design on the left

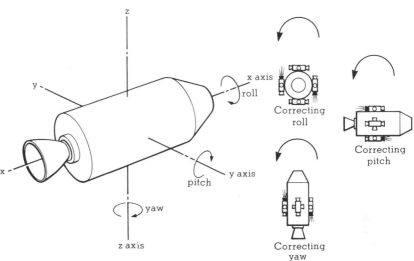

Correcting roll

Correcting pitch

Correcting yaw

Favoured propellants for the thrusters are the simplest ones — monopropellants such as hydrazine or nitrogen gas. Standardized refills of these two propellants (right) be can now be obtained at most interplanetary bases; they are available in re-usable spherical titanium tanks.

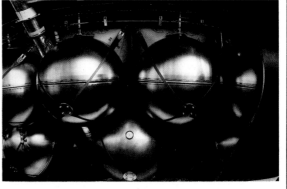

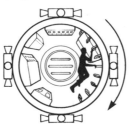

ATTITUDE CORRECTION IN AN EMERGENCY

Even if all the thruster propellant has been used up, it is sometimes possible, depending on the type of spacecraft, for the astronaut to turn the craft by himself. If there are convenient handholds around, say, the inside circumference of the ship, the astronaut could pull himself rapidly around, to make the spacecraft rotate in the opposite direction. This is really only practicable in small ships, as the spacecraft reacts to only a small fraction of the astronaut's efforts.

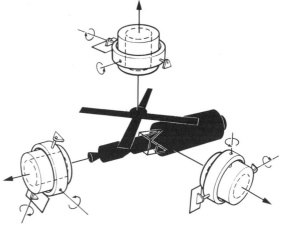

GYROSCOPIC STABILIZATION

Heavy-duty stabilizing gyroscopes can be used to counter any movement in a spacecraft almost instantaneously. First used in Skylab, their only disadvantage is that friction can cause erratic behaviour.

Guidance systems Interplanetary mapping

HAVING ESTABLISHED a stable orientation for the spacecraft, the astro-navigator's next task is to monitor continuously the ship's location in space and its velocity. There are several equipment systems that he can use to do this, and in practice they function automatically, for the most part. Guidance of the spacecraft is so closely interrelated with attitude control that it is not surprising to find equipment such as the star-tracker doing both jobs.

Navigation in space means two things — finding the position, which can be done by visual star tracking or by radio tracking, and determining the velocity, which can be done by radio tracking or with an inertial instrument such as the accelerometer.

Star-tracking sensors

The same sensors that help the spacecraft to keep in the right attitude also help locate its position, in the same way that a sailor takes bearings to fix his position. The angles are measured between reference stars, such as Canopus and the Sun, and, in combination with radio signals from Earth and interplanetary bases, the computer fixes the position.

Radio tracking, antenna

Radio beacons are now well established throughout the Inner System and at key locations in Jupiter Space and Saturn Space. They transmit easily identifiable signals in known frequencies, and all the relevant information about them, including orbits, is readily available on tapes for the ship's computer. The astronavigator uses his directional antenna to obtain bearings, and these neatly complement visual star-tracking.

On the major Solar System bases, such as Clavius Base on the Moon, and Chryse Base on Mars, as well

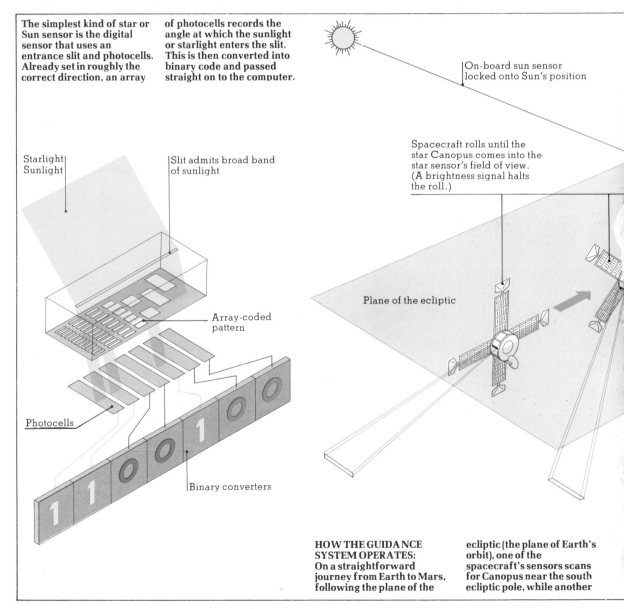

The simplest kind of star or Sun sensor is the digital sensor that uses an entrance slit and photocells. Already set in roughly the correct direction, an array of photocells records the angle at which the sunlight or starlight enters the slit. This is then converted into binary code and passed straight on to the computer.

Starlight
Sunlight

Slit admits broad band of sunlight

Array-coded pattern

Photocells

Binary converters

On-board sun sensor locked onto Sun's position

Spacecraft rolls until the star Canopus comes into the star sensor's field of view. (A brightness signal halts the roll.)

Plane of the ecliptic

HOW THE GUIDANCE SYSTEM OPERATES: On a straightforward journey from Earth to Mars, following the plane of the ecliptic (the plane of Earth's orbit), one of the spacecraft's sensors scans for Canopus near the south ecliptic pole, while another as at several locations on Earth, major tracking facilities are available. Their large antennae and computers have a phenomenal accuracy, and under certain conditions can temporarily take over the guidance of small craft. Except in an emergency, this service has to be paid for, of course, and this can add to the cost of the voyage by an appreciable amount. Most spacecraft captains rely on their own navigational systems.

A different function of the radio antenna is to measure velocity, and this can be done with surprising accuracy. The way it is done is by measuring the Doppler shift in signals. As a vehicle moves away from, say, a ground-based antenna, its radio transmissions change in frequency. In other words, each successive signal takes longer to travel from the transmitter to the receiver. This can easily be measured, and even on the very first unmanned inter-

planetary probes velocity could be measured to an accuracy of 0.00001 per cent. A typical system in practice would use an S-band transponder, which accepts a signal at one known frequency, then alters it proportionately to re-broadcast it. In this way, using frequencies that are known before the start of the journey, the Doppler shift can be measured both ways.

Accelerometers

These measure the acceleration of the spacecraft quite independently of outside help, and so are very flexible, if less accurate than the Doppler method. The simplest illustration of an accelerometer is a weight attached to a spring and lying on a flat surface. Any acceleration along the line of the spring would compress or extend it. In a spacecraft's accelerometer, a magnetic coil replaces the spring, but the principle remains the same.

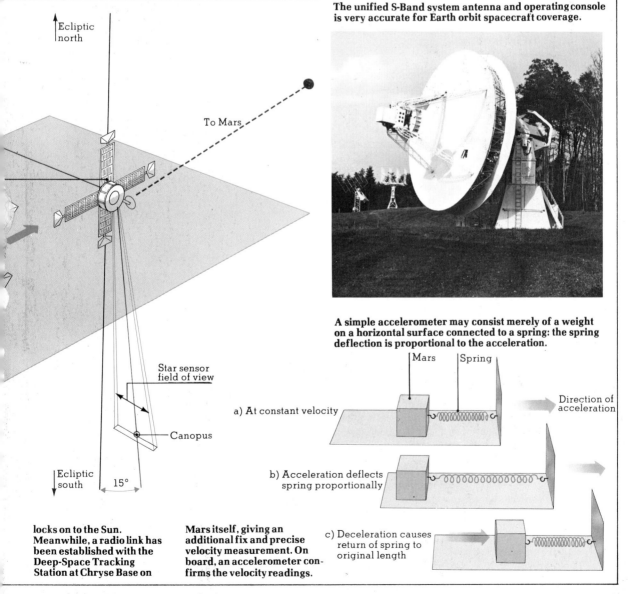

The unified S-Band system antenna and operating console is very accurate for Earth orbit spacecraft coverage.

A simple accelerometer may consist merely of a weight on a horizontal surface connected to a spring: the spring deflection is proportional to the acceleration.

a) At constant velocity

b) Acceleration deflects spring proportionally

c) Deceleration causes return of spring to original length

Ecliptic north

To Mars

Star sensor field of view

Canopus

Ecliptic south   15°

locks on to the Sun. Meanwhile, a radio link has been established with the Deep-Space Tracking Station at Chryse Base on

Mars itself, giving an additional fix and precise velocity measurement. On board, an accelerometer confirms the velocity readings.

Observation Key to survival

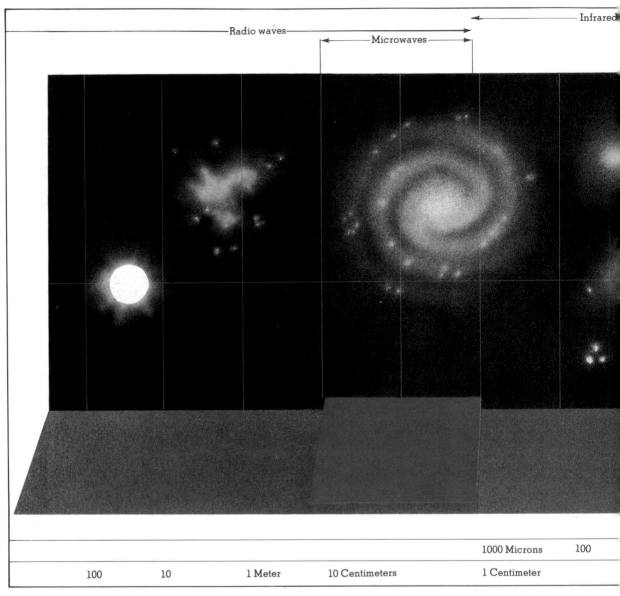

Radio waves

Microwaves

Infrared

| 100 | 10 | 1 Meter | 10 Centimeters | 1 Centimeter |

1000 Microns   100

SPACE IS A COMPLEX and hostile environment, and charting its courses demands a massive and sophisticated input of information. Without the screening effect of Earth's atmosphere, the range of information available is enormous, ranging from 1 000 meter radio waves to gamma rays measuring less than one angstrom — and this is just the electromagnetic spectrum! As a result, observational equipment is key to the spacecraft's survival, and a large proportion of a ship's expenditure and weight is devoted to these instruments. Inexperienced crew members often view data acquisition as tedious and irrelevant, but the inevitable astronavigational emergency normally rids them of this attitude.

Instruments may either be integrated with the ship's hull, being installed during construction, or if a variety of missions is likely to be undertaken during the ship's life, they may be grouped on a science-scan platform.

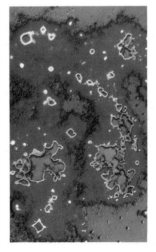

**Left: nebular emission in the Magellanic cloud, a nearby galaxy, photographed in far ultra- violet light (1216 Angstrom) with a UV camera. The galaxy is a 'young' one, consisting of 100 million stars.
Right: Voyager 1's science-scan platform, used for collecting and transmitting data from Jupiter and Saturn
Top: The electromagnetic spectrum describes the range of wavelengths of the protons which travel through space at the speed of light. The wavelengths vary from radio waves, thousands of meters long, to gamma rays, measured in ten-millionths of millimeters, or Angstroms.**

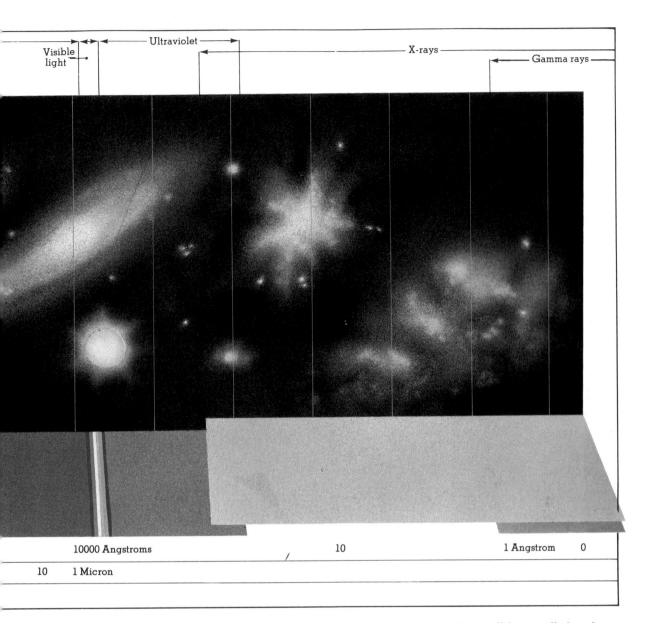

Visible light

Ultraviolet

X-rays

Gamma rays

10000 Angstroms   10   1 Angstrom   0

10   1 Micron

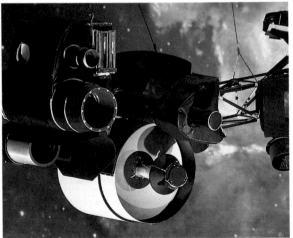

In any case, some of them will be installed on boom arms to isolate them from the influence of the ship's power systems.

Electromagnetic Sensing Instruments
By far the most important range of equipment. Not all may be found on one spacecraft.

Imaging Sciences Subsystem — the Vidicon Tube
Although photographic instruments, using normal emulsions, are sometimes used as ship's equipment, they cannot be linked to the computer's data-processing subsystem as easily as a direct scanner, which records visual information automatically. Optical systems are interchangeable, although it is normal to use both a wide-angle lens and a catadioptric mirror leescope. If the optical axes of both are bore-sighted, simultaneous and compatible

images can be obtained. Internal filter wheels permit colour recording and polarized pictures, whilst optional time-lapse equipment permits controlled sequences.

The sensitive vidicon faceplate records the image, which is then scanned into a sequence of 800 lines each with 800 picture elements (pixels). Each pixel is assigned one of 28 intensity levels, and all this information is recorded digitally on tape. It can, of course, also be monitored in real time. Resolution is excellent, and at 1000 km from the surface of a planet, the mirror lens can resolve down to 19 meters.

Infra-red sensing

Infra-red investigation is of extreme importance when approaching any planetary body with an atmosphere. Atmospheric gases have various absorption bands throughout the infra-red, and valuable information about their composition and thermal structure, as well as optical properties, can be obtained with the normal infra-red apparatus. This traditionally comprises a telescope for collecting the information, two interferometers for spectra in the near and far infra-red, and a radiometer for measuring total reflection.

Photopolarimetry

The Sun radiates unpolarized light — that is, in waves that vibrate equally in an infinite number of planes. When this light is scattered by a surface or an atmosphere, the reflected light ends up vibrating more in some planes than in others. It is this light-scattering effect of surfaces that the photopolarimeter measures, by taking readings through three polarizing filters set 60° apart. By taking a set of readings at different angles and under different conditions of illumination, a surprising amount of information can be gleaned about a planet's surface. The photopolarimeter can distinguish, for example, between bare rock, dust, frost, ice and gravel. In addition, it is indispensable for close approaches to Saturn's rings, as it can determine the type of particles and measure their size to a high degree of accuracy.

Ultra-violet Spectroscopy

Solar UV radiation and the normal traffic of energetic particles stimulates the atmospheric gases of planets to produce an 'airglow', that can be measured by a spectrometer sensitive to the right portion of the band.

The normal UV spectrometer is sensitive to the far end of the band, 500 A to 1700 A. It is quite useful for voyages in Jupiter space for detecting the variable sodium and hydrogen clouds that precede and follow Io's orbit around the parent planet.

The UV spectrometer has two modes — airglow, when it must be as sensitive as possible, and occultation. Here it is pointed directly at the Sun to measure a planet's atmosphere through examining how it blocks the waves entering it. When the instrument is mounted on a science-scan platform, a mirror is used for the occultation mode, so that the other instruments are not damaged by having them point directly at the Sun.

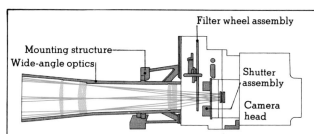

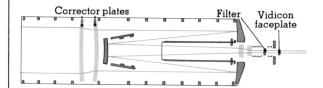

Imaging science instruments: the wide-angle camera (top) and the narrow-angle camera (bottom) function by resolving scattered light.

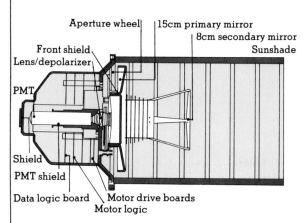

Configuration of the photopolarimeter which provides information about light-scattering surfaces or atmospheric particles.

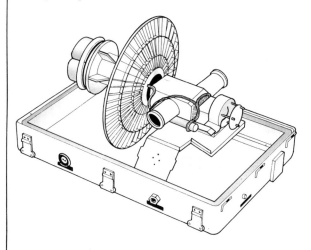

The infrared interferometer spectrometer is used in the study of atmospheres, and of the heat balances of the outer planets.

The ultraviolet spectrometer is used in the investigation of the make-up and density of the atmospheres of the outer planets and their moons.

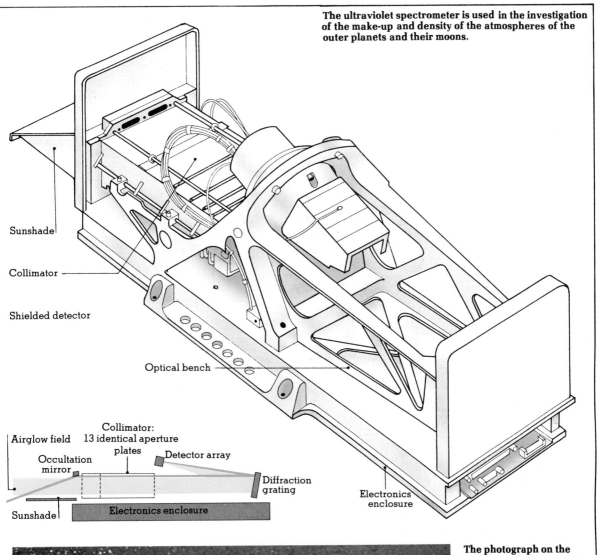

Sunshade

Collimator

Shielded detector

Optical bench

Electronics enclosure

Airglow field

Occultation mirror

Collimator: 13 identical aperture plates

Detector array

Diffraction grating

Electronics enclosure

Sunshade

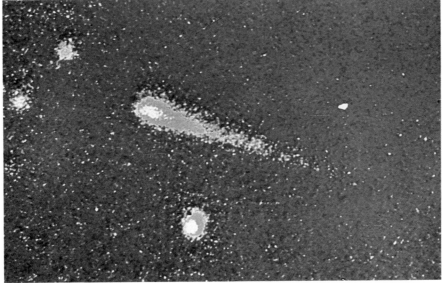

The photograph on the left is of the comet Kohoutek. It passed Earth at Christmas in 1973, and the photograph was taken from Skylab. Most of the billions of comets travel in the remote gravitational fr nges of the Solar System.

133

Observation Standard methods

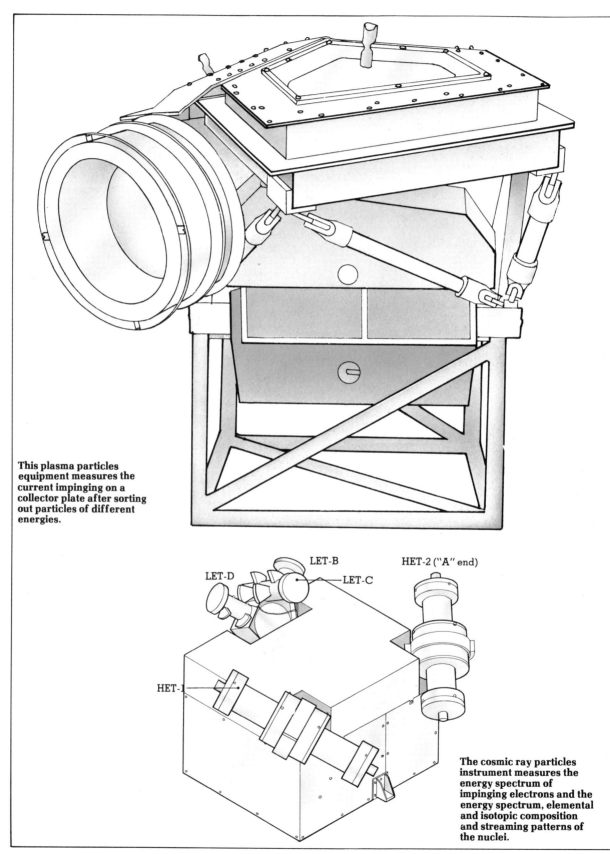

This plasma particles
equipment measures the
current impinging on a
collector plate after sorting
out particles of different
energies.

LET-D   LET-B   LET-C   HET-2 ("A" end)

HET-1

The cosmic ray particles
instrument measures the
energy spectrum of
impinging electrons and the
energy spectrum, elemental
and isotopic composition
and streaming patterns of
the nuclei.

Radio Sciences Subsystem — S-band and X-band transmitters/receivers

Strictly speaking, the S-band and X-band radio equipment (where both are carried) are part of the Communications System, but, with careful monitoring at a receiving station, they can be used for more occultation experiments. Two spacecraft acting together can be particularly effective in measuring such things as the refraction of an atmosphere. Of more immediate practical use, radio tracking of the spacecraft either from a planetary station or another ship can reveal unevenness in a planet's gravitational field. The small deviations of a craft in low orbit can show up the existence of mascons (large areas where the density of the planet's crust is greater).

Cosmic Rays

Cosmic rays originate from a number of sources outside the Solar System, and consist of particles travelling at close to the speed of light. Most are protons, but some of them are electrons or stripped nuclei. Because of their high velocities, they pack enormous energy and are therefore a hazard. As they are modulated by solar activity, a constant check must be kept on them. A typical cosmic ray particles instrument is a package combining a high- and low-energy telescope system (HETS and LETS), and a separate electron telescope (TET).

Plasma Particles

Plasmas are gases composed of charged particles and, for the astronaut, the two most important plasma fields in the Solar System are the solar wind and the magnetosphere of Jupiter – other planets, of course, have magnetospheres, but Jupiter has the most potent. Interaction between the solar wind and the Jovian magnetosphere makes an on-board detector indispensable for up-to-the-minute readings. The normal sensor consists of two Faraday cups aligned along different axes, one of them always pointing in the direction of the solar wind.

Magnetic Fields

Another crucial measurement of the Jovian (and other) magnetospheres is made with a magnetometer package, invariably supported at some distance from the spacecraft on a boom, to avoid interference from the ship's systems. Both high-field and low-field magnetometers are normally employed (HFM and LFM), each with sensors pointing along three axes for directional readings.

Radio Astronomy

Interstellar radio astronomy is, of course, the preserve of the major planet-bound telescopes, and is of little practical interest to the run-of-the-mill spacecraft. Nevertheless, a limited form of planetary radio astronomy is sometimes practised, normally using antennae no more than 10 meters. Receivers for both low-frequency and high-frequency bands are usually employed, and once again these are of greatest use in the vicinity of Jupiter, which emits strong signals to its close satellites in a complex relationship.

A Deep Space Network 64 meter radio antenna

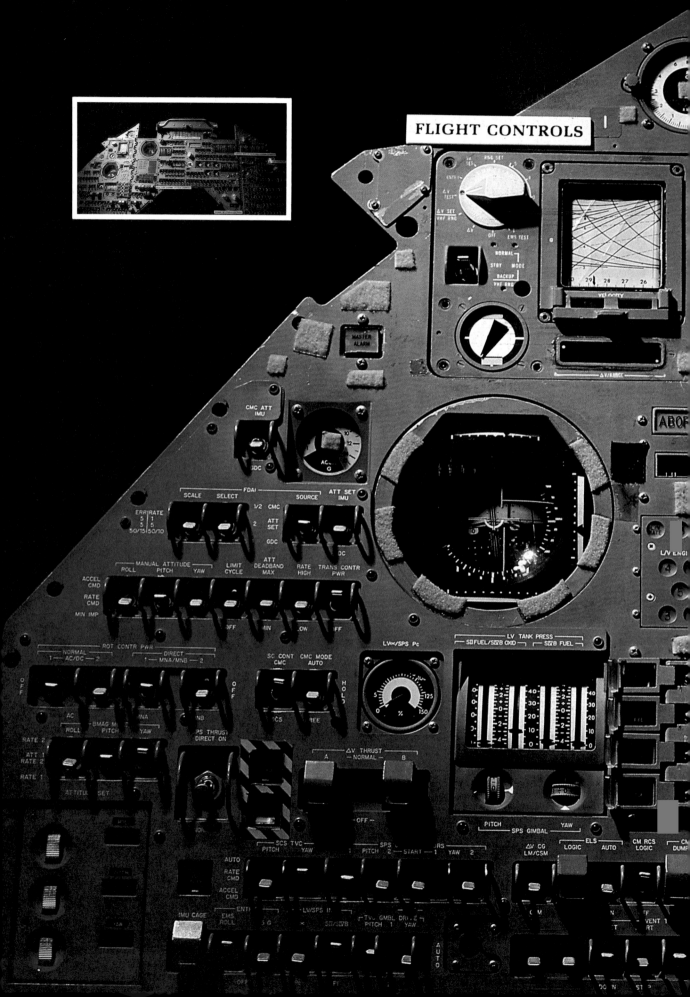

FLIGHT CONTROLS

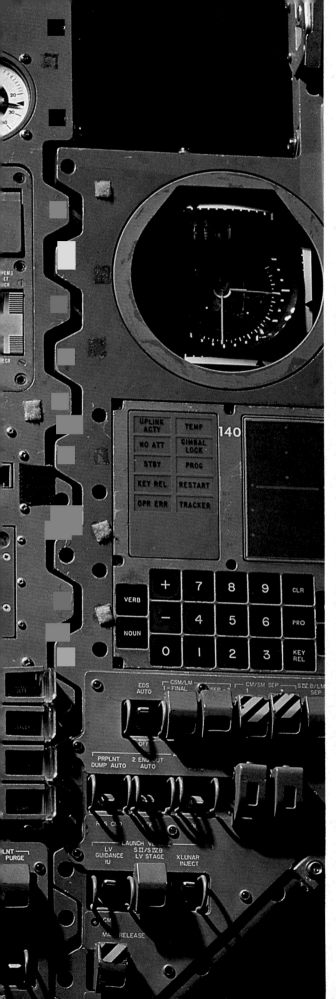

Control panels

Many readers will be familiar with the control panels shown here of (left) the Apollo command module and (bottom) the Apollo lunar module. Both are popular attractions at space familiarization centres, where eager prospective travellers are encouraged to understand and operate, in simulated circumstances, many of the controls. While these exercises are conducted in a fairly light-hearted fashion, tutors find that most people respond with child-like excitement to the opportunity to feel what it must have been like to be first on the Moon, and their understanding of modern systems certainly benefits. The Shuttle Orbiter panel (below) is less popular, but equally instructive.

Field Guide Technology to date

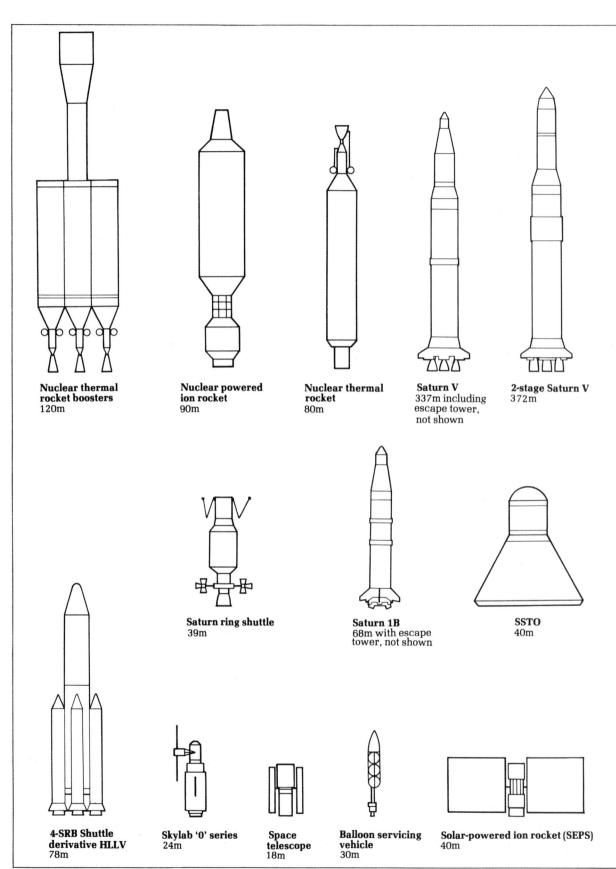

Nuclear thermal rocket boosters
120m

Nuclear powered ion rocket
90m

Nuclear thermal rocket
80m

Saturn V
337m including escape tower, not shown

2-stage Saturn V
372m

Saturn ring shuttle
39m

Saturn 1B
68m with escape tower, not shown

SSTO
40m

4-SRB Shuttle derivative HLLV
78m

Skylab '0' series
24m

Space telescope
18m

Balloon servicing vehicle
30m

Solar-powered ion rocket (SEPS)
40m

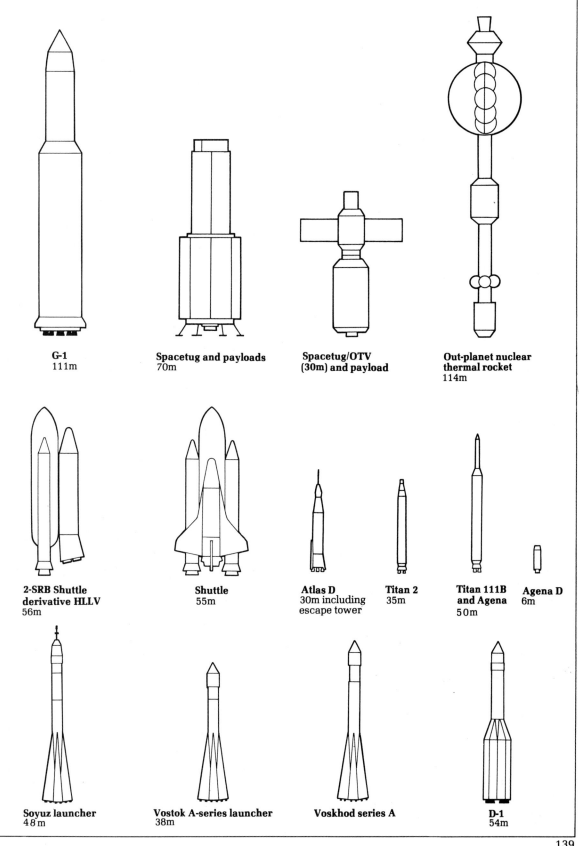

G-1
111m

Spacetug and payloads
70m

Spacetug/OTV
(30m) and payload

Out-planet nuclear
thermal rocket
114m

2-SRB Shuttle
derivative HLLV
56m

Shuttle
55m

Atlas D
30m including
escape tower

Titan 2
35m

Titan 111B
and Agena
50m

Agena D
6m

Soyuz launcher
48m

Vostok A-series launcher
38m

Voskhod series A

D-1
54m

Field Guide Technology to date

Lunar one-man flyer
3m

Mars trike
1.4m

Mars open one-man rover
3m

Mercury shielded rover
3.4m

LRV
3m

Moon rover enclosed

Lunar one-man flying unit

Lunar sub-orbital one-man flyer
2m

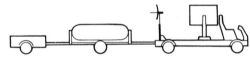

Moon rover with transporter and trailers
4.2m

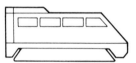

Moon rover-bus enclosed
5.5m

One-man work module
3m

Manipulator module
3.6m

One-man work module
2.5m

Belt miner
20m

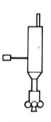

Shielded Jupiter orbiter
15m

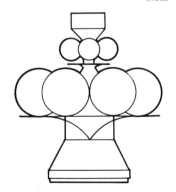

Daedalus
220m

Out-planet H/HE transporter
200m

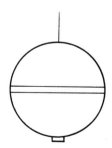

Belt refinery
146m

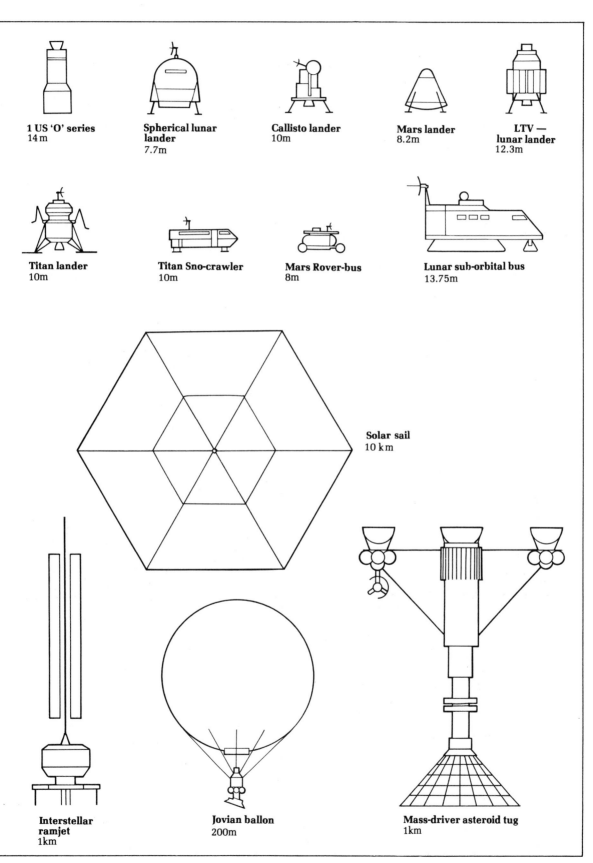

1 US 'O' series
14 m

Spherical lunar lander
7.7m

Callisto lander
10m

Mars lander
8.2m

LTV — lunar lander
12.3m

Titan lander
10m

Titan Sno-crawler
10m

Mars Rover-bus
8m

Lunar sub-orbital bus
13.75m

Solar sail
10 km

Interstellar ramjet
1km

Jovian ballon
200m

Mass-driver asteroid tug
1km

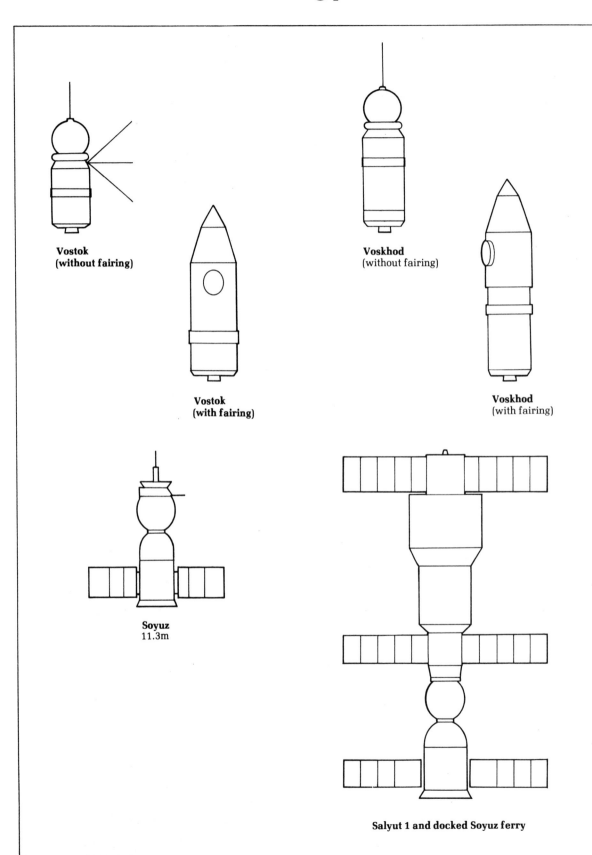

**Vostok
(without fairing)**

**Vostok
(with fairing)**

Voskhod
(without fairing)

Voskhod
(with fairing)

Soyuz
11.3m

Salyut 1 and docked Soyuz ferry

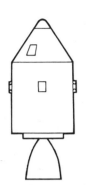

Apollo command module
2.6m

**Apollo command/
service module**
10m

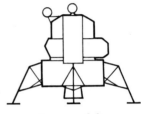

Gemini capsule
6.5m

Mercury capsule
3.3m

A.T.D.A.
3.5m

Lunar module
6m

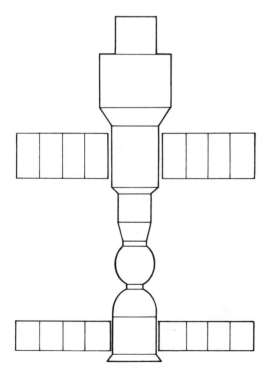

Salyut 3 and docked Soyuz ferry

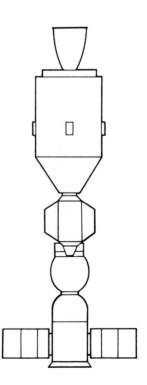

ASTP (Apollo/Soyuz link-up)

Geography of space

WHEN, IN 1596, Sir Walter Raleigh wrote 'The Discovery of the Large, Rich and Beautiful Empire of Guiana', he wrote of an Arcadian New World, and captured the romance of exploration in its golden age. Now, sad to say, the Earth is all too familiar to most people; even the Amazon forests and the Himalayas offer few opportunities to experience the excitement of discovery. Within the last century, however, we have been able to reach not just one New World, but many. Even a journey to Mars has all the elements of real travel, while the Out-Planets offer a challenge to the most intrepid explorer.

A research ship inserts into a
parking orbit over Callisto,
the closest safe moon to
Jupiter. Observational
instruments on boom arms
have been mapping Jupiter's
energy output.

The Solar System An overview

IN THE BOONDOCKS of the Milky Way, itself only a middle-sized galaxy, an unremarkable yellow dwarf supports a normal‾ retinue of planets. Our Sun, one of 100,000 million other stars in the Galaxy, is in the Carina-Cygnus spiral arm, 27,000 light years from the Galactic Centre.

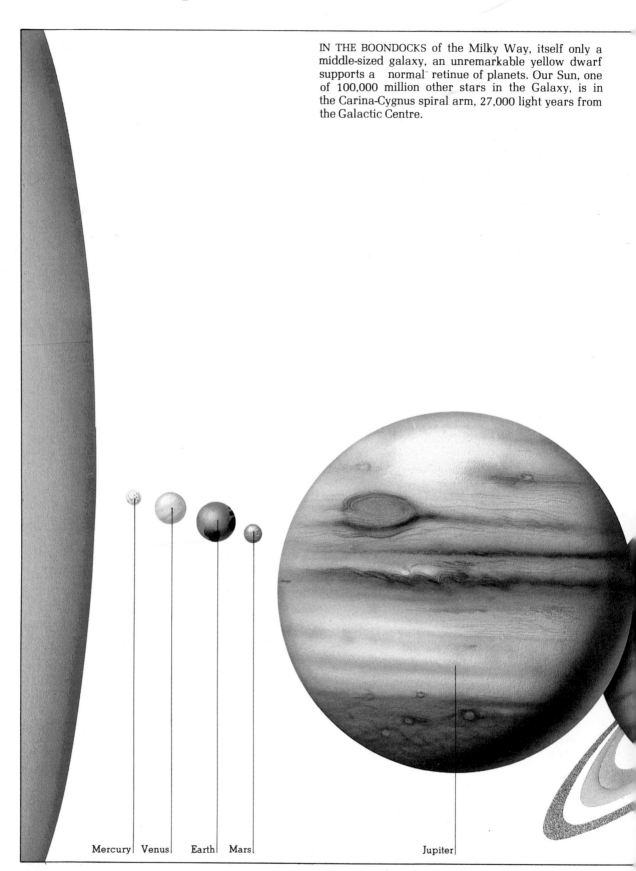

Mercury Venus Earth Mars Jupiter

The position of our Sun and its System in our galaxy, the Milky Way, is indicated by the arrow

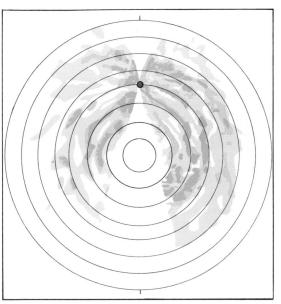

The above map shows the distribution of hydrogen atoms in our galaxy. It shows a distribution ranging from 0.05 atoms per cubic centimeter for the lightest tint to 1.6 for the darkest. The distribution of the atoms closely parallels the distribution of young stars.

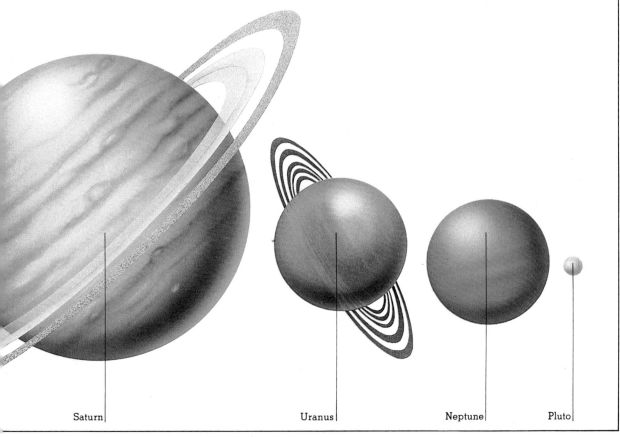

Saturn

Uranus

Neptune

Pluto

The Solar System Basic facts

While away an interplanetary hour browsing over the vital statistics of the planets of our Solar System, here presented in tabular form for easy comparison.

	MERCURY	VENUS	EARTH	MARS
Maximum distance from the Sun (Millions of kilometers)	69.7	109	152.1	249.1
Minimum distance from the Sun (Millions of kilometers)	45.9	107.4	147.1	206.7
Mean distance from the Sun (Millions of kilometers)	57.9	108.2	149.6	227.9
Mean distance from the Sun (Astronomical units)	.387	.723	1	1524
Period of revolution around Sun	88 days	224.7 days	365.26 days	687 days
Period of rotation on axis	59 days	—243 days (Retrograde)	23 hours 56 minutes 4 seconds	24 hours 37 minutes 23 seconds
Orbital velocity (Kilometers per second)	47.9	35	29.8	24.1
Inclination of axis to orbit	<28°	3°	23° 27′	23°59′
Inclination of orbit to ecliptic	7°	3.4°	0°	1.9°
Eccentricity of orbit	.206	.007	.017	.093
Equatorial diameter (Kilometers)	4880	12,104	12,756	6787
Mass (Earth = 1)	.055	.815	1	.108
Volume (Earth = 1)	.06	.88	1	.15
Density (Water = 1)	5.4	5.2	5.5	3.9
Oblateness	0	0	.003	.009
Main components of atmosphere	None	carbon dioxide	nitrogen, oxygen	carbon dioxide, argon (?)
Mean surface temperature (Degrees Celsius) (S = solid, C = clouds)	35 (S) Day —170 (S) Night	—33 (C) 480 (S)	22 (S)	—23 (S)
Atmospheric pressure at surface (Millibars)	10^{-9}	90,000	1000	6
Surface gravity (Earth = 1)	.37	.88	1	.38
Mean apparent diameter of the Sun	1°22′ 40″	44′ 15″	31′ 59″	21′

JUPITER	SATURN	URANUS	NEPTUNE	PLUTO
815.7	1507	3004	4537	7375
740.9	1347	2735	4456	4425
778.3	1427	2869.6	4496.6	5900
5203	9539	19.18	30.06	3944
11.86 years	29.46 years	84.01 years	164.8 years	247.7 years
9 hours 50 minutes 30 seconds	10 hours 14 minutes	—11 hours retrograde	16 hours	6 days 9 hours
13.1	9.6	6.8	5.4	4.7
3°05′	26°44′	82°5′	28°48′	?
1.3°	2.5°	.8°	1.8°	17.2°
.048	.056	.047	.009	.25
142,800	120,000	51,800	49,500	6000 (?)
317.9	95.2	14.6	17.2	.1 (?)
1316	755	67	57	.1 (?)
1.3	.7	1.2	1.7	?
.06	.1	.06	.02	?
hydrogen, helium	hydrogen, helium	hydrogen, helium, methane	hydrogen, helium, methane	none detected
—150 (C)	—180 (C)	—210 (C)	—220 (C)	—230 (?)
—	—	—	—	—
2.64	1.15	1.17	1.18	?
6′ 09″	3′ 22″	1′ 41″	1′04″	49″

The Solar System Seven wonders

THE ORIGINAL Seven Wonders of the World were first selected by Antipater of Sidon in the 2nd century B C He deemed them essential viewing for sightseers in the Alexandrian World. Now that the traveller's horizons have been extended almost to the limits of the Solar System, it is high time that a new selection be made, one that looks further than the Egyptian pyramids, the Colossus at Rhodes, and the Hanging Gardens of Babylon. Unfortunately, the works of Man can hardly compare with the truly monumental natural wonders of the planets, and some of the Wonders that the editors have now selected are not only a marvellous and awe-inspiring spectacle, they are also a challenge to the understanding of the casual

1. SOLAR FLARES
A complex pattern of magnetic fields shapes and controls the magnificent prominences that frequently erupt from the Sun's surface. A simple coronagraph is sufficient to appreciate the delicate traceries and elegant arches of a typical flare, large enough to dwarf the Earth.

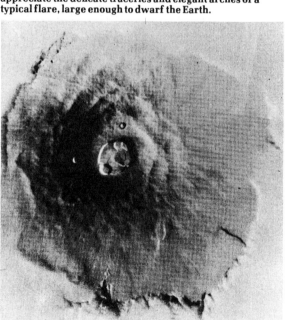

2. OLYMPUS MONS
The mighty caldera (collapsed crater) of Olympus Mons measures 80 km across, at a height of 24 km above the surrounding plain. The base measures 600 km in diameter and is marked by an enormous cliff edge so that, from orbit, the volcano appears to sit on the planet's surface. In spring and summer, afternoon clouds cloak the western slopes as the air rises towards the summit; visible even from the Earth, these clouds suggested the name Nix Olympica (Snows of Olympus) to Earthbound astronomers.

3. MERCUREAN DOUBLE SUNRISE
Due to the combined effect of a fairly eccentric orbit and a day that is in resonance with its year, Mercury displays the unique sight of a double sunrise once in every revolution about the Sun. The mechanics of this remarkable phenomenon are fully described on p. An experience for connoisseurs of the unusual, the double sunrise can be appreciated every 88 Earth-days at either 90° or 270° longitude.

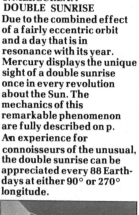

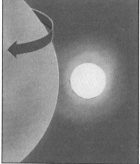

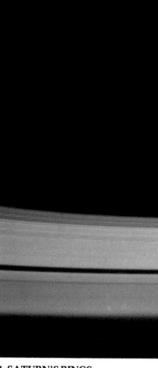

4. SATURN'S RINGS
A breath-taking spectacle that displays elegance and symmetry from every angle, Saturn's ring system has been the subject of countless artistic interpretations. Most highly recommended.

5. TITAN
Titan's varied landscape and changeable atmospheric conditions provide a welcome relief from the monotony of the airless cratered satellites that are all too common throughout the system. On a fine day, with russet clouds scudding across the face of Saturn, hanging low in the blue sky, Titan can seem almost homely to the visitors.

visitor as they once were to the professional scientist.

One or two of the Wonders almost select themselves, such as the rings of Saturn and Jupiter's Red Spot, while the inclusion of others is unavoidably arbitrary. In the case of Olympus Mons, the giant Martian volcano, some consideration was given to the larger Venusian contender; in the end, however, the sheer

inaccessibility of the surface of Venus when set against the majestic setting of Olympus Mons, tipped the scale in favour of Mars. In any event, sheer size is hardly sufficient qualification in itself.

Objectively, Earth must be considered the jewel of the System, but as this guide is for its jaded inhabitants, Terran wonders have not been included.

6. MARINER VALLEY
Left: surprisingly unknown until the voyage of Mariner 9, this Martian version of the Grand Canyon dwarfs its Arizonan counterpart. Embracing 3700 km of the Martian equator, the rift — actually a series of inter-connecting canyons — measures up to 240 km across with a depth of, in places, nearly 7 000 meters. This is four times that of the Grand Canyon.

7. JUPITER'S RED SPOT
40,000 km long by 11,000 km wide, the great Red Spot in Jupiter's south tropical zone is the Solar System's longest-running bad weather. It is an enormous hurricane tinged a delicate orange-red by hydrocarbon compounds. Lethal radiation from the planet makes the Red Spot best viewed from the safety of Callisto.

The Solar System Historic sites

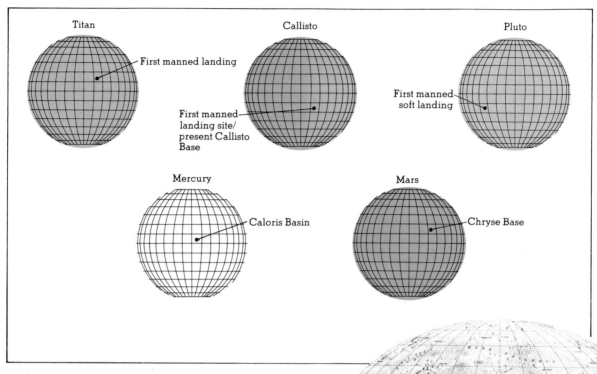

Titan — First manned landing

Callisto — First manned landing site/present Callisto Base

Pluto — First manned soft landing

Mercury — Caloris Basin

Mars — Chryse Base

AS PART OF THE FIFTY YEARS IN SPACE celebrations of 2011 the Space Heritage Institution and the Planetary Parks & Reservations Service were founded under the auspices of the ISA. Subsequent arguments over billboard rights at the Apollo 11 landing site have resulted in a temporary withdrawal of financial support.

Nevertheless, a schedule of Historic Sites and Planetary Parks has now been agreed, and some work has been completed on their protection. Best known, and most visited, of course, is the site of the first Moon landing. A protective compound has recently been sprayed over those footprints belonging to Armstrong and Aldrin that survived the take-off blast of their Lunar Module and subsequent vandalism, and visitors now tour the site by following a railed boardwalk.

Some of the later Apollo sites are more picturesque and relatively unspoiled, whilst the landing and impact sites of the unmanned craft such as Rangers, Surveyors and Lunokhods are likely to be of more interest to the enthusiast. The Moon's airless conditions are ideal for preserving even the finest detail of a footprint, and most visitors remark that the sites appear to have been abandoned only the day before.

Mars, in many ways, has more evocative sites. The first manned landing site is incorporated in Chryse Base and rather lacks atmosphere, but a visit to the grave of four members of the South Polar Expedition of 2014, who met their tragic end in the icy wastes of the Argyre Plain, is a powerful and emotional experience.

The administration of restricted areas is undertaken directly by the ISA. Beacons and precise navigational directions can be found in the relevant publications (e.g. the Belt Pilot pub. 2049 and revisions, the Jupiter Pilot pub. 2029 and revisions).

Above: the Apollo landing sites. Apollo 15 carried out the Moon programme's first 'j' mission of heavier scientific content. The Apollo 16 astronauts explored the Descartes region for over 20 hours. Apollo 17, in 1972, brought back a record 250 pounds of samples.
Right: the official first footprint on the Moon's surface. There are rival claims, which seem to fit Armstrong's shoe size more closely.
Inset: In November 1969 Apollo 12 landed in the Ocean of Storms, to find Surveyor 3, serene and silent against the horizon.

Hazards Major and minor

SPACE TRAVEL is not without a certain amount of danger. In conditions where life has to be supported completely artificially, small errors and minor equipment breakdowns can have fatal consequences. However, despite the inherent dangers, space is not as capricious as the seas or mountains of Earth. Most of the hazards of interplanetary travel can be measured, predicted and insured against.

The two hazards that have received most attention over the years are radiation and meteoroid collision, the one a very real danger, the other insignificant. Probably because we can see the potential danger just by looking at the Moon's craters, the danger of meteoroid collision has been a popular theme in science fiction, but in reality the risk is negligible. But radiation, from power supplies and engines as well as from flares and cosmic rays, is a major threat.

Space debris is a minor hazard. Since the first launch — Sputnik — in 1957, many thousands of spacecraft, manned and unmanned, have been put into space, most of them into Earth orbit. With the increase of launching agencies and the uncertainty of the life of orbiting hardware, it is now very difficult to say what is in space, let alone exactly where it all is.

The crater in the lunar surface (right) is 50 miles in diameter. This gives some idea of the destruction a meteorite can wreak when it does actually make impact.

IONIZING RADIATION — A REAL THREAT

Normally, the Sun issues a stream of particles into space in the form of the solar wind — between 5 and 10 protons, electrons and particles per cubic centimeter. Travelling at 500 km/sec, they cannot penetrate sufficiently to harm humans. However, during a violent solar disturbance, high energy protons are blasted into space with little warning. Astronauts may have only a few minutes to get under cover, and the dangerous proton flux can last for days after a solar flare. Galactic cosmic rays must always be shielded against. On the left is a helmet used for Apollo 8. It was riddled in space with cosmic rays making tracks half a millimeter long in the shape of the replicas shown.

Radiation doses are measured in rems. Most places on Earth have a background of 0.1 rem per year, and US standards for the general population are 0.5 rem per year. The annual dose from cosmic rays on an unshielded astronaut is a high 10 rems. So some form of shielding is needed all the time, and special temporary shields must be available for solar flares.

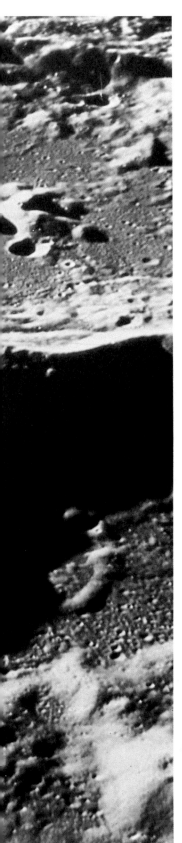

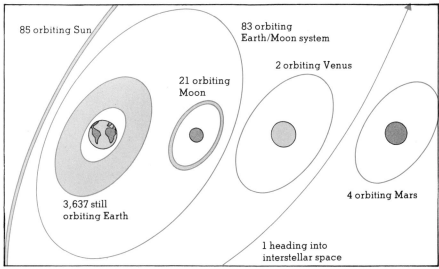

85 orbiting Sun

83 orbiting
Earth/Moon system

21 orbiting
Moon

2 orbiting Venus

4 orbiting Mars

3,637 still
orbiting Earth

1 heading into
interstellar space

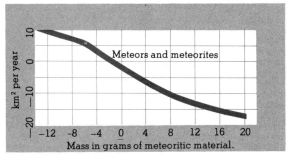

Meteors and meteorites

km² per year

Mass in grams of meteoritic material.

METEOROID COLLISION – A NEGLIGIBLE DANGER

The graph above shows the rate of collision on one square kilometer near the Earth in a single year. Information from all kinds of sources, from early Apollo missions to Earth-based observations, shows a consistent pattern: on average, a 1 gm meteoroid will cross a square kilometer of space every 10 years, a 100 gm meteoroid every 5 000 years, whilst a 10 kgm meteoroid would be expected only once every 100,000 years. Near the Earth most of these travel at a speed of about 40 km/sec relative to the Sun –

fast enough to do a lot of damage, but the danger of collision seems remote.

The only problem is that meteoroids are not evenly distributed in space, but come in showers, so that if and when a spacecraft does receive a hit, it may well be hit several times. Standing outside on a clear night, the observer on Earth may see an average of 10 meteors an hour on a normal night. During peak showers this rate can rise to 50 an hour. In 1833 over Paris during the Leonid shower (see below), up to 35,000 meteors an hour were seen.

SPACE DEBRIS — AN ORBITING SCRAPYARD

This picture, in 1975, shows the 3883 man-made objects remaining in space after 18 years of spaceflight. By this time, 5 000 had already landed or re-entered.

PROTECTING THE SPACECRAFT AGAINST METEOROIDS

If, by misfortune, a meteoroid does hit the spacecraft, the amount of damage it will do depends on its mass. To start with, *any* particle colliding at the typical speed of 40 km/sec will make a crater, but a surprisingly thin skin will prevent penetration. A typical manned inter-planetary craft has an area of about 1500 sq m, and with a skin just 5 mm thick there might be one penetration a year (by a small particle). Aluminium 7cm thick would cut the rate down to once every 10,000 years.

However, even better protection is afforded by the Whipple meteor-bumper, first proposed as early as 1952. A thin metal skin surrounds the main wall of the spacecraft at a short distance. This fragments meteoroids and dissipates much of their energy.

If, after all this, a respectably-sized meteoroid does get through the defences, then the effects can be unpleasant, especially for anyone close to the entry hole. A 10 gm meteoroid hitting a space-craft at 40 km/sec has the punch of 2 kgm of TNT.

155

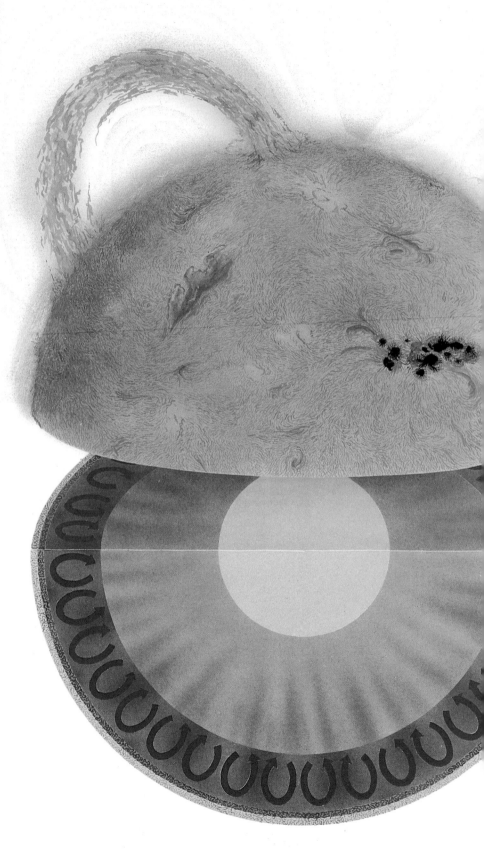

Cross-section of the sun
shows its most likely
structure. Proton-proton
chain reactions fuse
hydrogen into helium
releasing energy that slowly
rises to the 6000 °K surface.
Near the surface turbulent
convection churns the hot
gases. The interwoven
magnetic fields are
responsible for sunspots
and flares.

Solar space Heart of the system

ENERGY SOURCE for the Solar System, our Sun is classed as a G2 star — medium-sized, moderately hot, in the middle of its life. Orbiting the centre of the Galaxy once every 200 million years, the Sun radiates energy into space from thermonuclear reactions at its core. Shedding part of its outer atmosphere in a steady stream of protons and electrons, the Sun is the source of the solar wind that blows across the whole planetary system at 500 km/sec.

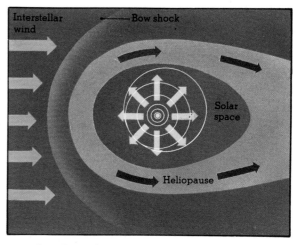

Few of our System's comets pass close enough to the Earth to display their 'tail', like Ikeya did (below, taken high in the Bolivian Andes). Tails can be so long they are measured in astronomical units.

Just as the Sun's solar wind blows gently across all the planets, so the interstellar wind blows around the entire Solar System on its 290 km/sec journey through Galactic space.

Solar space

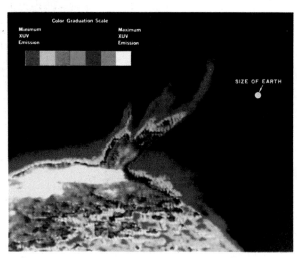

Color Graduation Scale

Minimum
XUV
Emission

Maximum
XUV
Emission

SIZE OF EARTH

PORTRAIT GALLERY OF THE SUN
Revealing different aspects of the Sun's character, a variety of imaging systems combine to give a comprehensive view of our nearest star.
Above:Colour-coded to show intensity, a major prominence erupts nearly one million kilometers into space.
Below: X-ray photograph of the Sun's magnetic fields, revealing hot bright magnetic loops and dark 'holes'.
Right: this false colour isophoto dramatically revealed a significant change in the coronal hole as compared to the following day. Solar rotation accounted for this. Skylab astronauts studied the Sun using this method.
Bottom: Skylab photographs a solar prominence leaping into space.

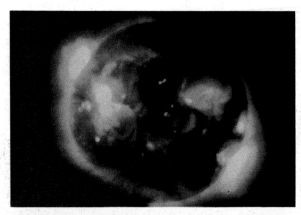

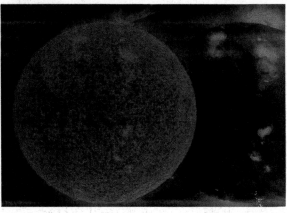

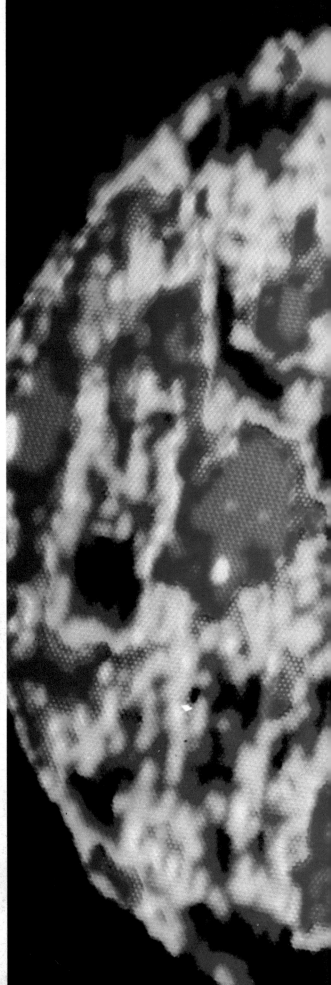

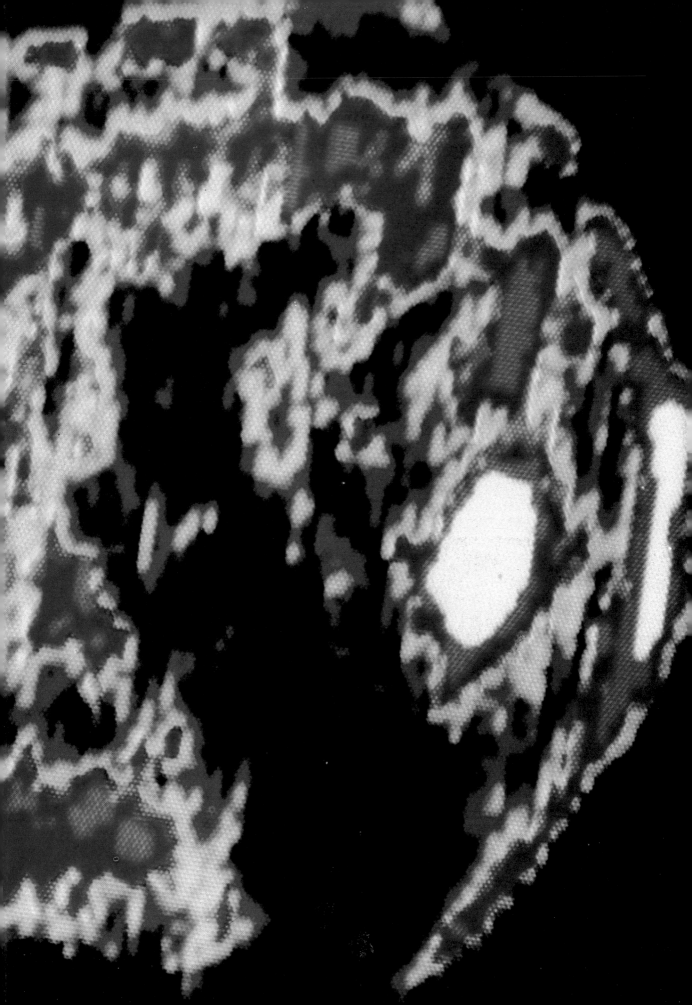

The planets Mercury

SMALLEST PLANET in the Solar System, Mercury's proximity to the Sun makes it one of the least attractive destinations. There is no permanent base, although an underground unmanned facility at the foot of the Schiaparelli Ridge is available for visiting research teams. For the visitors, Mercury inevitably invites comparison with the Moon, and the general terrain has much the same appearance, being divided between densely-cratered highlands and poorly-cratered plains (like the lunar maria). The striking Caloris Basin immediately reminds the arriving traveller of the Moon's Mare Orientale. Crater walls tend to be lower on Mercury, and are surrounded by less ejecta, due to the higher gravity (0.37 G compared with the Moon's 0.165 G).

Without a doubt, Mercury has an unappealing environment. At the closest approach to the Sun the noon-time temperature on the equator can reach 430°C, whilst at the other extreme the darkside temperature averages —160°C. There are seasons on Mercury, but unlike the Earth or Mars, they vary not with latitude but with longitude. This is because Mercury's rotation is locked to its orbit around the Sun. It rotates exactly three times for every two orbits, and as a result the 0° and 180° meridians receive two and a half times as much solar radiation as the 90° and 270° lines of longitude.

This 'spin-orbit coupling', as it is called, is responsible for one of the Solar System's more unusual sights. Because Mercury's orbit is quite eccentric, there is a short time during which the orbital speed overtakes the rotational speed, and over a period of eight days the visitor can watch the Sun slowly come to a halt in the sky, then move backwards before resuming its path towards the horizon! Admittedly, as this happens at Mercury's closest approach to the Sun, considerable fortitude and an efficient refrigeration system are required.

A SOLAR LOOP-THE-LOOP
High spot of the hardy traveller's itinerary is Mercury's once-every-orbit solar 'loop'. At perihelion (closest approach), the Sun goes backwards for a short while. Mercury's axis is nearly perpendicular to its orbital plane, so its seasons change with longitude.

The view (right) from an approaching spacecraft shows Mercury's western hemisphere, with the terminator crossing the striking Caloris Basin. One thousand three hundred kilometers across, this basin is surrounded by the 2,000 meter high ranges of the Caloris Mountains. Across from these mountains, large plains were formed by sheets of out-welling lava.

Above: close-up of part of the Caloris Basin, probably formed by the impact of a planetismal.

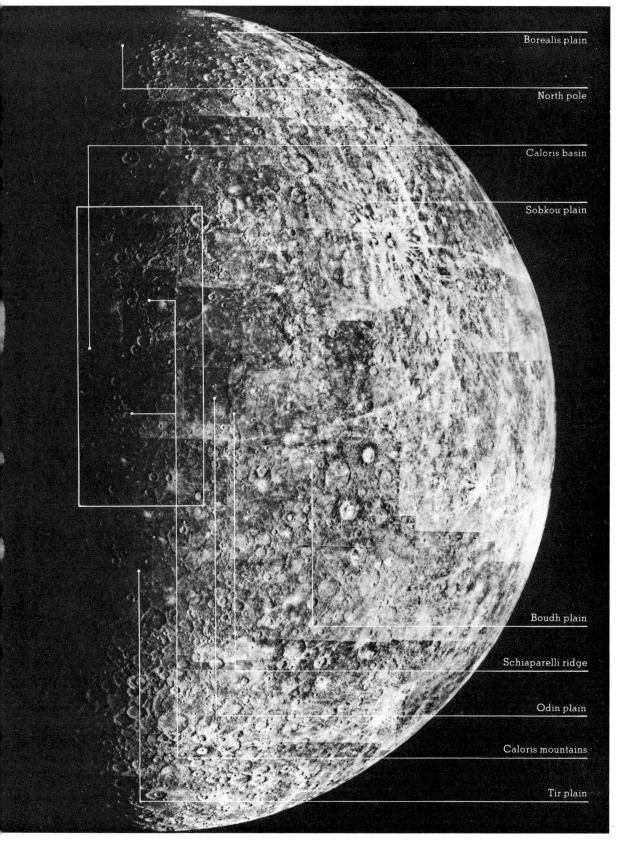

Borealis plain

North pole

Caloris basin

Sobkou plain

Boudh plain

Schiaparelli ridge

Odin plain

Caloris mountains

Tir plain

The Planets Venus

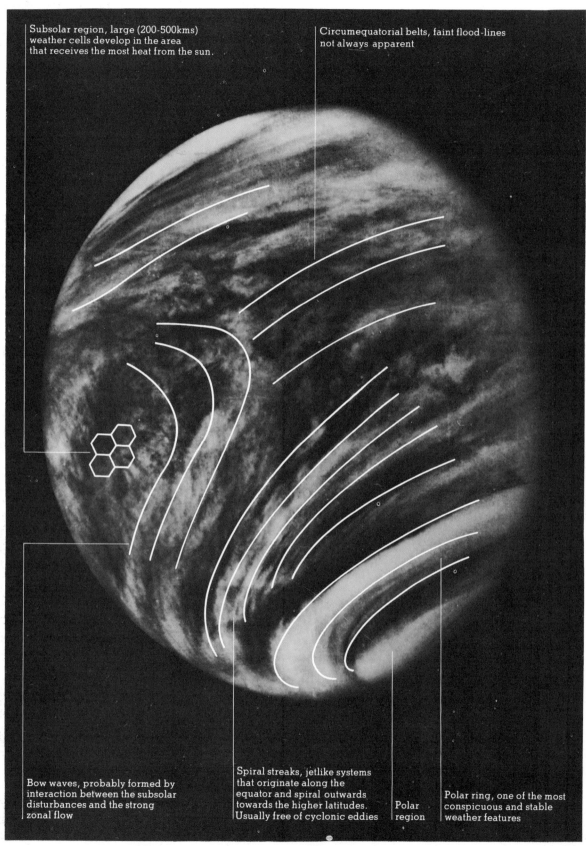

Subsolar region, large (200-500kms)
weather cells develop in the area
that receives the most heat from the sun.

Circumequatorial belts, faint flood-lines
not always apparent

Bow waves, probably formed by
interaction between the subsolar
disturbances and the strong
zonal flow

Spiral streaks, jetlike systems
that originate along the
equator and spiral outwards
towards the higher latitudes.
Usually free of cyclonic eddies

Polar
region

Polar ring, one of the most
conspicuous and stable
weather features

SO SIMILAR TO THE EARTH in mass, density, diameter and its position in the Solar System, it is hard to imagine a planet more inimical to Earth life. Being closer to the Sun than the Earth, Venus naturally receives more energy — almost twice as much — but this alone does not explain the high temperatures, extreme pressures and corrosive atmosphere at the surface. What has happened is a runaway 'greenhouse effect'. The dense atmosphere of carbon dioxide allows sunlight through to heat the surface, but traps outgoing heat radiation, resulting in surface temperatures of around 470°C and pressures equivalent to those at one kilometer under the Earth's oceans. Possibly the extra solar radiation that Venus receives was enough to trigger off the process by releasing carbon dioxide from the rocks.

Venus rotates very slowly — once every 243 days — and in the opposite direction to that of the Earth. This slow rotation may be the explanation for the planet's virtual lack of magnetic field.

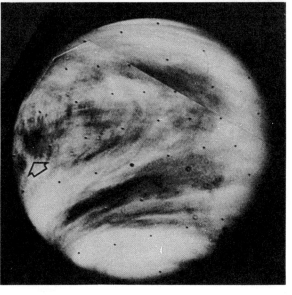

A METEOROLOGIST'S DELIGHT
Left: Venus displays its high-altitude weather in ultra-violet light. Atmospheric motion is characterized by broad strokes, uncomplicated by the local distrubances that make analysis of the Earth's weather so difficult. The basic recurring pattern is a large horizontal 'Y' centred on the equator.

HIGH WINDS ON A SLOW-MOVING PLANET
Although the planet itself rotates only once every 243 days, cloud features in the stratosphere, about 100 kilometers above the surface, swirl around Venus at high speeds. The sequence on the right was taken at seven hour intervals, showing that some features can travel right round the planet in four days.

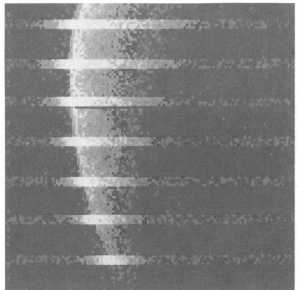

ULTRA-VIOLET PHOTOGRAPHY
This form of photography has been essential to our understanding of Venus' weather. Vidicon cameras, such as this (used on Mariner 10 in 1970), have an extended ultra-violet response, and record the image from high-quality mirror optics onto a television frame of 700 scan lines, each sampled 832 times. The resolution of this camera was good enough to pick out features 65 km across at a distance of over three million km.

The planets Venus

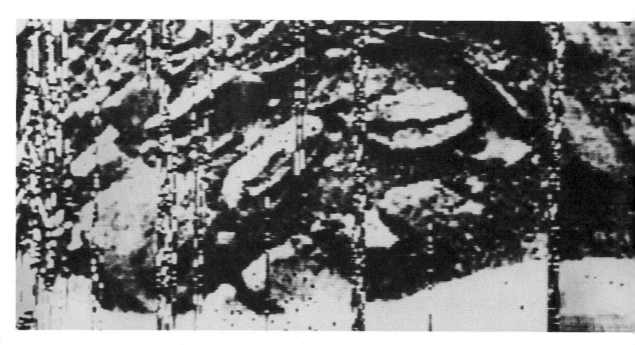

A MODEL FOR DANTE'S INFERNO
Above: the surface of Venus presents conditions so extreme that manned landings are out of the question, unless the climate can be altered (see 'A New Earth?'). This photograph is the first ever taken on the surface, by the Russian lander Venera 9 in 1975. Although the winds at this level are light and balmy, they blow in temperatures of 470°C in an atmosphere of 97 per cent carbon dioxide and less than one per cent water vapour, 90 times as dense as that of the Earth. Only a spacecraft with the structure of a bathyscape could survive. The spacecraft that took this picture survived only 53 minutes, which was longer than its makers hoped for.

ON A CLEAR DAY YOU CAN SEE FOREVER
Right: a more recent photograph, taken in 1998, shows a Venusian panorama. The refractivity of the enormously dense atmosphere is so great that the light is bent. If the air were clear enough, the view would be similar to that at the bottom of a fishbowl, and the whole planet would appear to curve upwards and outwards from the observer. In reality, light-scattering and clouds restrict visibility; the surface is bathed in a ruddy gloom.

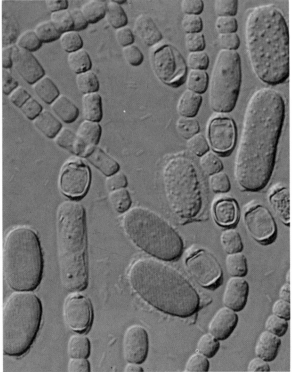

The Solar Wind reacts with Venus in a much less spectacular way than with most other planets, because, for all practical purposes, Venus has no magnetic field (less than 1/10,000 of Earth's).

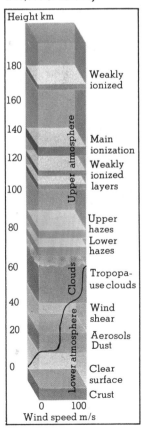

A NEW EARTH?
First suggested in 1961 by Carl Sagan, terraforming is now being given serious attention by planetary engineers, who may have devised a way to create conditions tolerable to man. Cancelling the 'greenhouse effect' that has turned Venus' atmosphere into a high-pressure furnace could be achieved by reducing the carbon dioxide content. This might be accomplished by seeding the atmosphere with algae, which could undertake the chemical engineering necessary to consume the CO_2 and produce O_2. Laboratory experiments with strains of blue-green algae such as *Cyanidum caldarium* have shown that this is a distinct possibility.

Earth Space Traffic problems at home

EARTH SPACE, which of course includes the Moon, is now relatively well-used, and navigation is increasingly subject to control and necessary restrictions. All travel within Earth Space falls under the jurisdiction of the Earth Space Control Net (ESCN), with which body all flight plans must be filed. Special qualifications are demanded of all astronauts using Earth Space routes, and, indeed, within the crowded zone of parking orbits, all craft must avail themselves of a Pilot. Naturally, unlike their maritime equivalents, space Pilots do not actually board the vehicles. Instead, they work from Ground Controls and are linked directly to the spacecraft's control and guidance systems via the network of Ground Tracking Stations.

Earth-to-orbit routes are the preserve of franchised Shuttle lines, but Moon landings and take-offs are regularly carried out by deep-space Pilots. Landing and service fees are applicable at all lunar bases.

The pattern of transportation in Earth Space is characteristically broken into definite 'legs'. Shuttle lines service the Low Earth Orbits (LEOs) from the Earth, while onward travel to the coveted and limited Geosynchronous Earth Orbit (GEO) is undertaken by the ubiquitous space tugs. Comsats, powersats and transfersats, as the communications, solar array and regular space stations are commonly known, vie for position in these stable locations. The tugs also transfer goods and passengers all the way to the Moon and *its* lunasynchronous orbits, as

well as to the Lagrange
Colonies.

Several companies now
offer popular and relatively
inexpensive 'lunar fly-bys',
lasting 12 days (289 hours),
passing close to the Moon's
surface and swinging past to
a wide apogee.
However, many travellers
will feel that, when they
have come so far, it is a
shame not to land and
savour the delights of low-G
sports, driving, and so on.
Indeed, although the Moon
could not yet be called the
honeymoon hotel of the
twenty-first century, its
popularity with new lovers
with large pockets is
growing all the time.

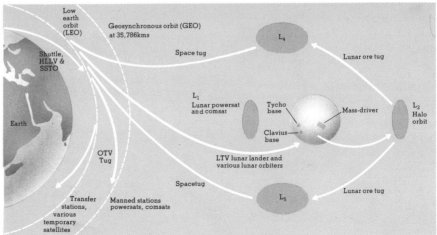

Earth Space Lunar living

THE FIRST CONTINUOUSLY — manned lunar base was established in 1994 at Clavius, a decade or two later than early NASA optimists had hoped. A rapid pace of construction somewhat compensated for this, and by 2008 the bases had achieved the effective status of colonies, with permanent populations.

In the construction of the bases, atmospheric pressure and radiation have been the major stumbling blocks, and have largely determined their appearance. From the arriving space tug, the visitor is immediately struck by the streamlined configuration of the domes and tubular modules. These rounded shapes help to maintain the atmospheric pressure in most parts of the base at near-Earth levels (1kg/sq cm) by efficiently resisting the various stresses. In addition, as the bases evolved in the earliest days, discarded spacecraft

parts, which tend to be cylindrical, were used as much as possible.

Of course, parts of the bases, notably the living quarters, are underground, protected from the cosmic background radiation and occasional flares by at least 5 meters of lunar basalt. It would indeed have been feasible to construct a completely subterranean base, but for psychological reasons it was felt to be important to have as much as possible above ground, with many viewing windows.

Although mercifully it has never happened, a 'blow-out' is the most feared accident, and the stringent safety precautions, involving alarm systems, quick-sealing emergency packs at regular intervals along the walls and numerous airlocks, reflect the concern at the back of every base-dweller's mind. For this reason, the bases

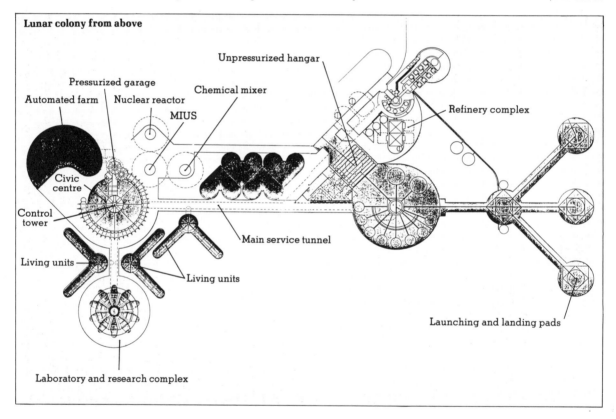

Lunar colony from above

Automated farm

Pressurized garage

Nuclear reactor

MIUS

Chemical mixer

Unpressurized hangar

Refinery complex

Civic centre

Control tower

Living units

Living units

Main service tunnel

Launching and landing pads

Laboratory and research complex

Left: Clavius base photographed from an approaching lunar lander. It may have the appearance of an airport to the first-time visitor,

Facing page, top right: an excellent view of a near full Moon, looking westerly towards the Sea of Crises on the horizon. East of Crises is the Border Sea, and, south of this the more circular area is Smyth's sea. Most of the lunar area shown here is on the farside. Far right: the dark patch is Crises. Near right: the Crater Tsiolkovsky.

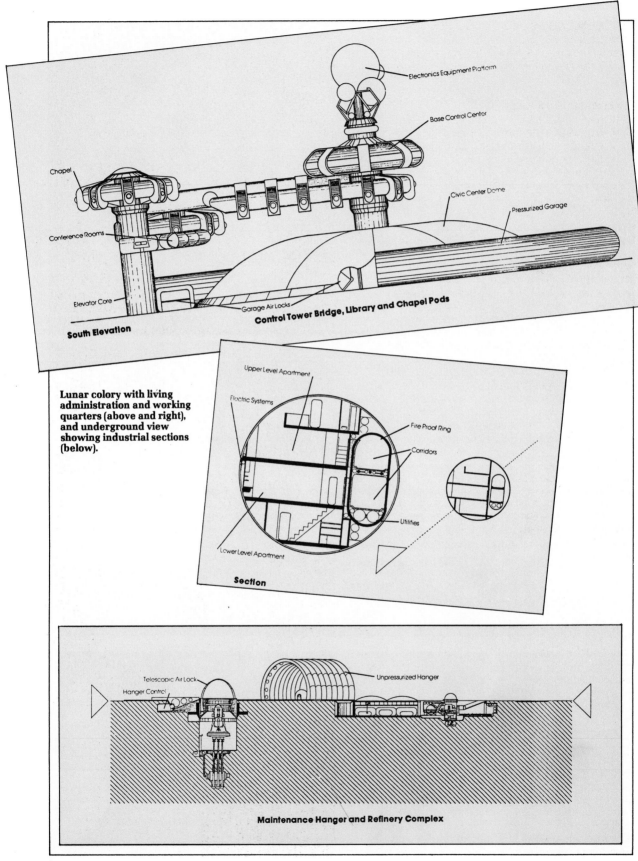

Electronics Equipment Platform

Base Control Center

Chapel

Civic Center Dome

Pressurized Garage

Conference Rooms

Elevator Core

Garage Air Locks

Control Tower Bridge, Library and Chapel Pods

South Elevation

Lunar colory with living
administration and working
quarters (above and right),
and underground view
showing industrial sections
(below).

Upper Level Apartment

Electric Systems

Fire Proof Ring

Corridors

Utilities

Lower Level Apartment

Section

Telescopic Air Lock

Unpressurized Hanger

Hanger Control

Maintenance Hanger and Refinery Complex

tend to be strung out, all the different functions decentralized and capable of being isolated from each other by the airlock system. In a way, each base can be considered as one long corridor, with the various units plugged into it.

Power and Utility System

At the heart of the lunar colony's survival is its power plant, a nuclear Thermionic Reactor with a capacity of 10 megawatts, sufficient to maintain an Earth city of 75,000 people. Solar power is used for various ancillary tasks, but the sheer efficiency of reactors makes them the mainstay of base power requirements. For safety the Reactor is buried in a 20 meter-deep well. Next to it is the Chemical Mixer, producing water and various compounds, and the Utility System, which manages the supply of water, air, waste and power.

Space Port and Maintenance

These landing pads accept incoming tugs, which are then moved by track to a fuel/cargo management area. Servicing is available in an underground hangar reached by means of a hydraulic elevator. Spacecraft supplies and fitting shops are nearby, as is a static engine testing facility. Lunar ground vehicles are garaged in this area.

Refinery and Manufacturing

This industrial complex processes rocks and minerals from nearby lunar mines. An electrolytic plant, catalytic cracker and a smelter are used to produce oxygen, water and building materials.

Farm Complex

After early preparation of an area of lunar soil with nitrogen, earthworms and other techniques, the farms are mainly devoted to fast-growing, protein-rich soybeans. All the base personnel's protein requirements are met by the main automated farm area, which has a 0.2 kilo/sq cm atmosphere of carbon dioxide and nitrogen.

Poultry and fish are reared in the heart-shaped experimental farms. Whilst less efficient than soybeans as a food, they are great morale-boosters.

Civic Center and Control Tower

The above-ground Civic Center is the base's social focal point and the key to the psychological success of lunar settlements. It contains swimming pools, restaurants and recreation areas (in high-jumping contests, 10 meters is a meagre achievement!). All this is encompassed in a nine-storey dome, held in place at the centre by the Control Tower, the operation nerve centre that overlooks the whole base area.

Living Accommodation

Tubular burrows below the lunar soil provide privacy and safety for the inhabitants of the base. Two-level apartments provide dwelling space for 240 people; power comes from the banks of solar panels on the surface. A hotel beneath the Civic Centre provides accommodation of a high standard for tourists and other temporary visitors.

Mare Serenitatis
Copernicus
Mare Imbrium
N
MareTranquillitatis
Mare Crisium

Tycho base
Clavius base
S
Mass-driver site
Apollo II landing site

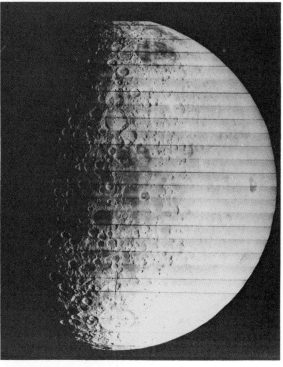

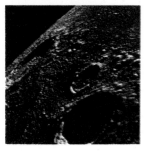

Above: the farside of the Moon, showing an earthquake-like fault at the lower centre near the south pole. The trough is 150 miles long and 5 miles wide. The centre of the photograph is about half way between the lunar equator and the pole. Left: the craters Copernicus (in the background) and Reinhold. Top: the most important lunar locations.

The planets Mars

AFTER THE 60 MILLION km, nine-month voyage from Earth, the approach to the red planet is breath-taking. On the final landing run, the craft passes over the Tharsis Ridge and its conspicuous volcanoes. Far to the south the 'Grand Canyon' of Mars is clearly visible.

Weather
Apart from the periodic duststorms that blow out of the south when Mars is at its closest to the Sun, the weather is quite consistent. Conditions are ideal for sightseeing: a typical weather report for Chryse Base might run as follows: light easterly winds in the late afternoon, becoming southeasterly after midnight; sky clear and cloudless; temperature minus 30°C in mid-afternoon falling to minus 85°C just before dawn; pressure 7.70 millibars.

Top: a computer-generated false-colour exaggeration of color variation on Mars, which corresponds to variation in clouds, atmospheric haze, surface frosts and rock materials.
Left: a mosaic of more than 1500 computer-corrected television pictures from Mariner 9 in 1971. The residual north pole ice cap is at the top.
Above: the rock-strewn surface of Utopia Planitia, seen for the first time via the cameras of Viking 2.
Right: on its approach to the dawn side of Mars in August 1976, the same craft captured Ascraeus Mons (top) with water ice cloud plumes on its western flank; the great rift canyon, Mariner Valley (middle); and (near the bottom) the large, frosty crater basin of Argyre.

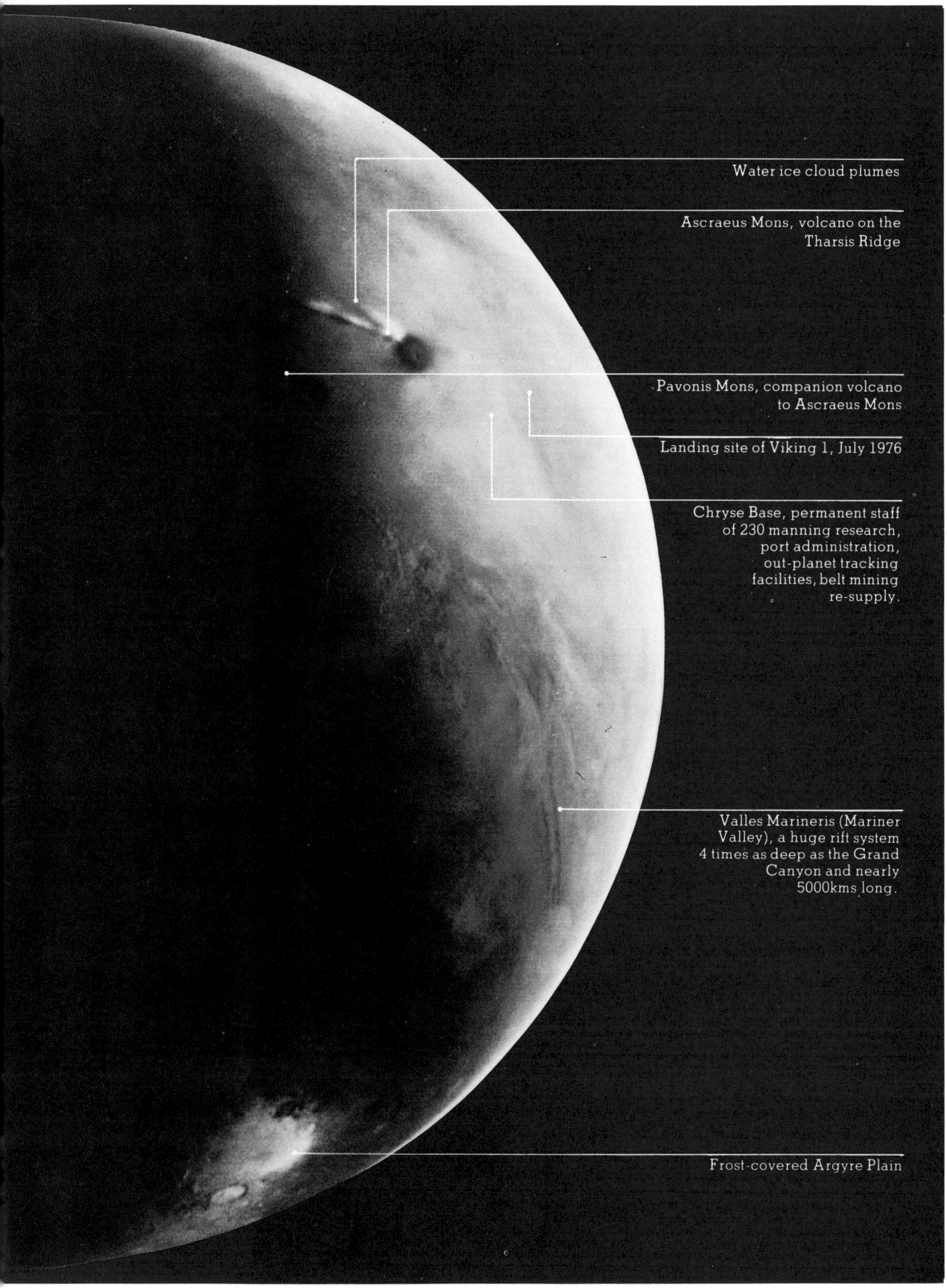

Water ice cloud plumes

Ascraeus Mons, volcano on the
Tharsis Ridge

Pavonis Mons, companion volcano
to Ascraeus Mons

Landing site of Viking 1, July 1976

Chryse Base, permanent staff
of 230 manning research,
port administration,
out-planet tracking
facilities, belt mining
re-supply.

Valles Marineris (Mariner
Valley), a huge rift system
4 times as deep as the Grand
Canyon and nearly
5000kms long.

Frost-covered Argyre Plain

The Planets Mars

The likeliest Martian tourist spots are indicated on the map on the facing page, with photographs from each location inset. The subjects of the photographs are as follows, working in a clockwise direction:

Top left corner, Olympus Mons. This gigantic volcanic mountain was photographed as the great Martian dust storm subsided. The clearing atmosphere revealed a mountain 500 kilometers across at the base. Steep cliffs dropped from the mountain flanks to a surrounding great plain. The main crater at the summit, a complex multiple volcanic vent, is 65 kilometers in diameter. The mountain is more than twice the breadth of the most massive volcanic pile on Earth.

Top right: visitors to Chryse Base will be interested to see a photograph which gave men their first idea of the appearance of the area. It was the first taken by Viking 1's camera on July 23 1976. The horizon is 3km away. The late afternoon Sun is in evidence on the left. The support struts on the left belong to the S-band antenna. The rocky surface shows wind-blown material and dunes. The nearest large rock is 3m in diameter and 8m from the spacecraft. The meteorology boom is right of centre. A cloud layer is visible half way between the horizon and the top of the picture. The sunset over Chryse Base (next) was taken four minutes after the Sun dipped below the horizon. Such spectacles serve to remove any doubts visitors might have as to the worthwhile nature of their trip. Argyre Plain (bottom right) is taken here when the sky was unusually clear. The layers of haze on the horizon are 25 to 40 kilometers high, and made of carbon dioxide crystals. The Mariner Valley (bottom left) is taken from a range of 4350 km. The two canyons running east-west are 60 km wide and more than half a mile deep. A few recent impact craters are bottom right.

Lastly, Pavonis Mons, rises more than 16 km above the surrounding plains. The caldera is a single large circular 48 km depression.

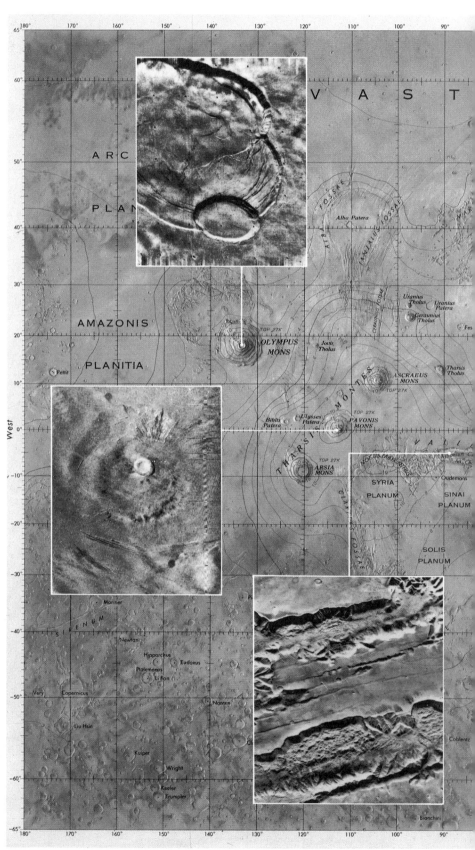

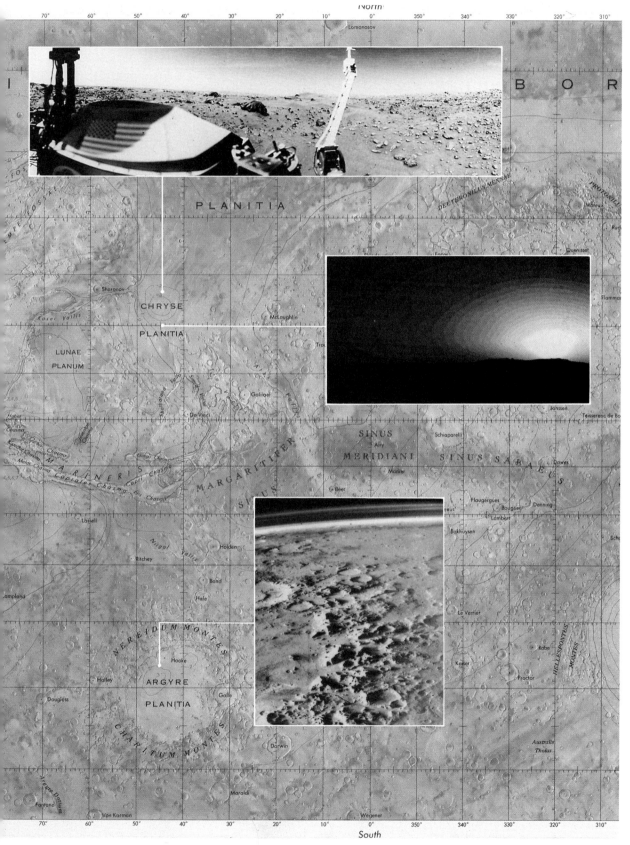

I

BOR

PLANITIA

CHRYSE

PLANITIA

LUNAE
PLANUM

Sharonov

Kasei Vallis

McLaughlin

Galilei

Da Vinci

MARINERIS

Ophir Chasma

Melas Chasma

Coprates Chasma

Capri Chasma

Eos Chasma

MARGARITIFER

SINUS

Lassell

Nirgal Vallis

Holden

Ritchey

Bond

Hale

NEREIDUM MONTES

Hooke

Halley

ARGYRE

PLANITIA

Galle

Douglass

CHARITUM MONTES

Argyre Dorsum

Fontana

Von Karman

Maraldi

Darwin

Wegener

Lomonosov

DEUTERONILUS MENSAE

PROTONILUS

Moreux

Quenisset

Flammarion

Janssen

Tisserenc de Bo

SINUS
MERIDIANI

Airy

Schiaparelli

SINUS SABAEUS

Mädler

Beer

Flaugergues

Bouguer

Denning

Lambert

Bakhuysen

Le Verrier

Rabe

HELLESPONTES MONTES

Kaiser

Proctor

Australis
Tholus

The planets Mars

REFLECTIONS OF PAST GLORY

In orbit over the planet, the visitor can see whole eras of Martian history. Ancient river beds and giant volcanoes recall a time of running water and a freshly created atmosphere. A rich pattern of long-dry river beds (right) cuts through the Lunae Planum.

Below: Olympus Mons thrusts its summit through the clouds to a height of over ten miles.

Facing page: early morning fog fills canyons in a high plateau in Labyrinthus Noctis. The clouds are tenuous and will soon be dispersed by the Sun. Now most of the Martian atmosphere is lost to space.

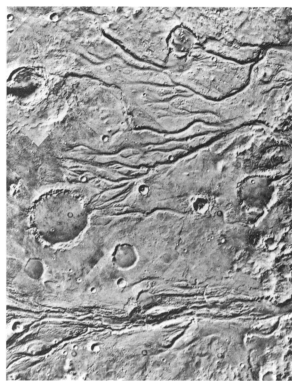

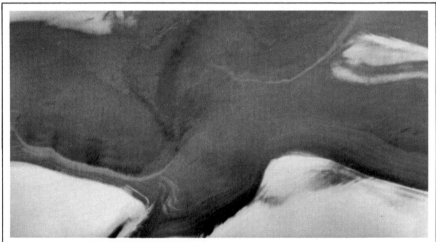

TERRAFORMING A LIFELESS PLANET

The North Polar expedition of 2013 confirmed the vast quantities of water locked in the ice caps: A controlled experiment has already begun to release this water into the atmosphere by sprinkling 100 sq km with carbon black. Eventually, it is hoped to treat the whole ice cap in this way; the carbon black will allow more of the Sun's heat to be absorbed, and the vaporization of the ice will increase the density of the Martian atmosphere.

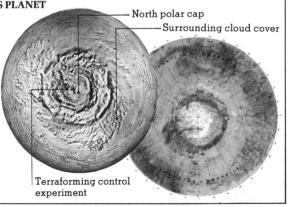

North polar cap

Surrounding cloud cover

Terraforming control experiment

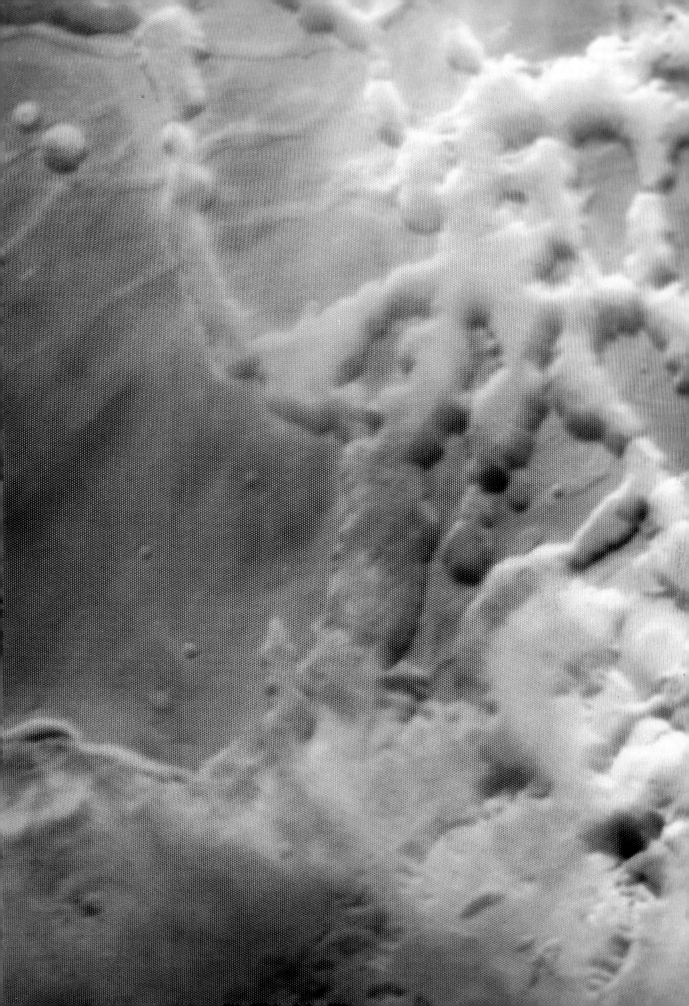

The Planets Mars

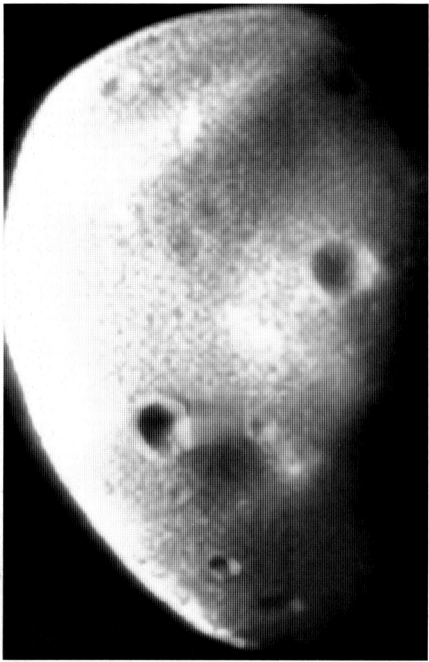

Deimos (above) is the outer
and smaller of the two
Martian moons. Its
dimensions are about 8 x 6 x
5 miles. Much of the
surface is covered in dust.
Facing page: Phobos, the
larger of the moons, and the
inner one. Its south pole
contains its largest crater.
This is named Hall, after its
discoverer. Phobos is so
close to Mars that it is under
considerable strain, and
may eventually break up.

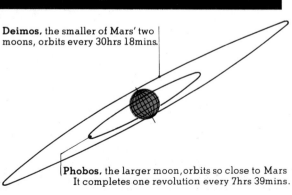

Deimos, the smaller of Mars' two
moons, orbits every 30hrs 18mins.

Phobos, the larger moon, orbits so close to Mars
It completes one revolution every 7hrs 39mins.

The asteroid belt

KNOWN THROUGHOUT the interplanetary mining community simply as 'The Belt', this huge ring of asteroids dominates space between Mars and Jupiter. Arguably the most significant sociological development in recent human history, the settlement of the Belt by independent miners and homesteaders is the spearhead of a movement to reassert the independence of the individual. The famous Belt Charter of 2045, signed only after two decades of skirmishing and illegal mining, opened the way to a new homesteading era.

Most Belters are fiercely independent, taking pride in their ability to survive virtually without Earth's assistance. There have already been booms and slumps, reminiscent of the gold rushes of the 19th century, for Belt mining is always at the mercy of constructional and technological demand on the Inner Planets. At present the Belt is in the grip of a petroleum boom, now that cheap and efficient zero-G refineries are available.

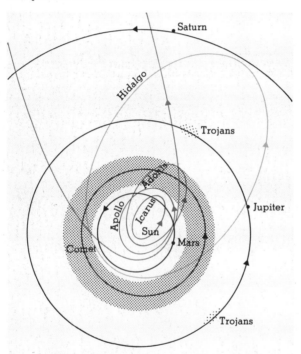

From near-planets to grains of dust, the asteroids (above) move in a wide variety of orbits, but most are concentrated in the 240 million km belt beyond Mars. Despite the numbers, and the size of some, not many are seen on the journey to Jupiter. Only about twenty house-sized chunks come within viewing distance on the 200 day transit of the Belt, and these must be watched for closely as most are very dark.

Right: a Belt mining and processing facility operated by a consortium closes with an ore-rich carbonaceous asteroid. Factories such as this are equipped for on-the-spot smelting and refining, but it is also common to tow complete asteroids to Earth space.

Most asteroids are carbonaceous chondrites, which contain almost half the organic-chemical content of oil shale, as well as water; others are of the stony-iron type, varying sometimes to pure nickel steel. Prospectors use spectroscopes to analyze their organic and mineral contents.

The planets Jupiter

ARRIVAL AT THE OUT-PLANETS after a two-and-a-half year voyage is a once-in-a-lifetime experience, and the attendant dangers can make this literally so. Jupiter, giant of the Solar System, must be treated with great respect by astronauts. A casual attitude towards astronavigation and radiation shielding will not serve the crew well.

A huge spinning ball of what is almost entirely liquid hydrogen, Jupiter emits twice as much energy as it receives from the Sun, and thus can rightly be considered a near-sun. Together with its fascinating variety of satellites (thirteen in all, more than any other planet), Jupiter virtually makes a mini-solar system.

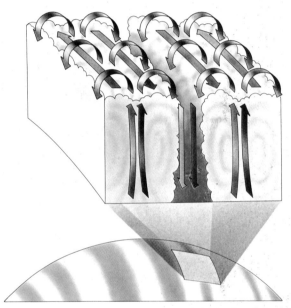

Europa (left) most of the planet is traversed by broad cloud bands. Jupiter's hot interior stirs these bands convectionally so that the light-coloured zones are rising gas and the dark belts descending gas. In the polar regions this banding is replaced by a turbulent mottling, indicating convectional weather cells. This process is shown in diagrammatic form above. The temperature range and chemical range on the planet are immense. Lower down in the atmosphere temperatures are between 35°C and 75°. Interestingly, the visible part of the atmosphere contains carbon-carrying methane.

Above the surface of liquid hydrogen, Jupiter has a thick, turbulent atmosphere, changing over a depth of 1000 km from water crystals at the base through ammonia crystals to cloud tops of gaseous hydrogen. Plainly seen from the gravelly surface of Europa, most of the planet is traversed by broad cloud bands. Jupiter's hot interior stirs these bands convectionally, so that the light-coloured zones are rising gas and the dark belts descending gas. In the polar regions this banding is replaced by a turbulent mottling, indicating convectional weather cells.

The planets Jupiter

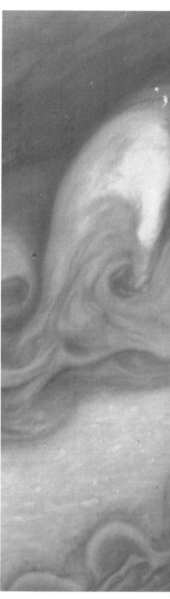

THE RED SPOT — STORM TO END ALL STORMS
Already listed as one of the sights of known space, this, the largest hurricane in the Solar System, rages over 40,000 sq km, towering 8 km over the surrounding clouds. At this altitude, phosphorus condenses to give the Red Spot its colour. The storm was first noted in 1664.
Left: Jupiter with three of its moons (Io, against it, the bright Europa and Callisto, bottom). The Red Spot is visible here and (below) from closer in. Right: it is flanked by a turbulent region to the west, indicated by the white ovals.
Facing page, bottom right: a multiple exposure gives the first evidence of the planet's thin ring.
Bottom left: the region to the east of the Red Spot.

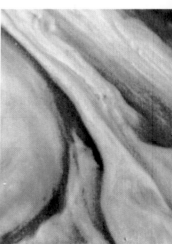

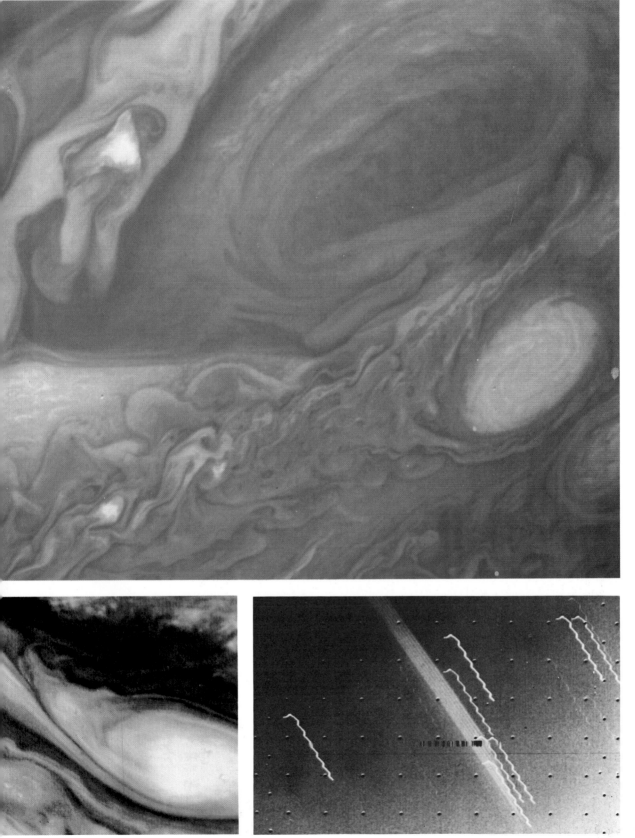

The planets Jupiter

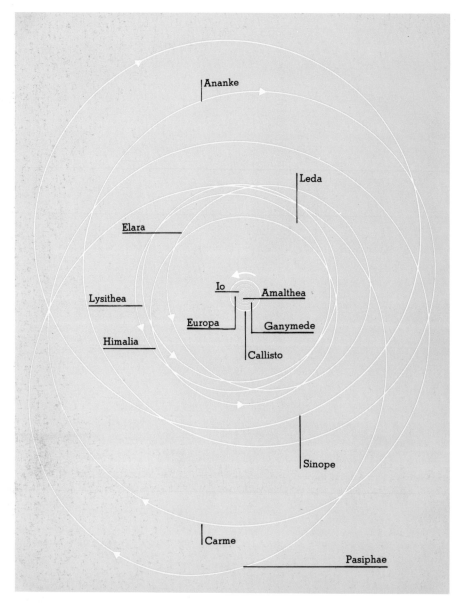

Ananke

Leda

Elara

Lysithea

Himalia

Io   Amalthea

Europa   Ganymede

Callisto

Sinope

Carme

Pasiphae

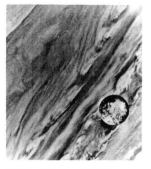

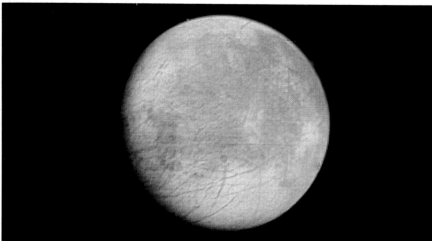

LETHAL RADIATION BATHES THE INNER MOONS.
Trapped by Jupiter's magnetic field and 10,000 times more intense than Earth's Van Allen Belts, radiation gives spectacular auroral displays. Travellers fortunate enough to survive a trip to the closest moon, Amalthea, which regularly receives 100 times the lethal radiation dose for humans, can testify to the awesome sight of the looming giant filling the sky.

Top: Io, the innermost moon, from 500,000 miles. The circular features may be a result of meteorite impact, or they may be of internal origin. The bright patches are younger deposits.
Above: Io against the parent planet. Bottom left: Europa, the smallest Galilean satellite, showing its unusual linear fractures, 1000 km in length and up to 300 km wide.
Right: Ganymede, half as big again as Earth's Moon and the largest of Jupiter's satellites. Only half as dense as our Moon, it is composed of rock and ice.
Bottom right: this photograph of Callisto shows features down to about 7 km across. Callisto, too, is made of ice and rock, its surface being 'dirty ice', or water – rich rock.

The planets Jupiter

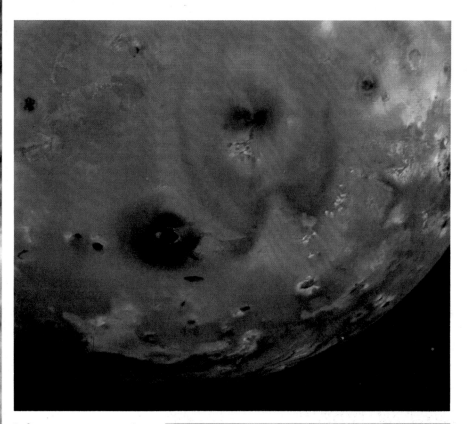

Left: an enormous vocanic explosion is silhouetted against dark space over Io. Solid material is thrown to a height of 100 miles, indicating an ejection velocity of 1200 mph. Below left: two of Jupiter's moons, Io and, on the edge of the picture, Europa, against a background of the Red Spot region. The photographs show the different colours of the two moons. Below right: Io from 77,000 miles. The reddish and orange colouring probably indicates sulphurous compounds and salts. The dark spot is a volcanic crater. Above: the relative lack of impact craters on Io suggests that its surface is younger than its neighbours.

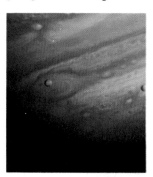

The planets Saturn

ONE THOUSAND AND FOUR HUNDRED MILLION kilometers from the Sun and a six year voyage on a Hohmann Transfer Orbit, Saturn is bathed in perpetual twilight. Nevertheless, the incomparable beauty of its rings makes it the jewel of the Solar System and high point on any tour. The rings of Uranus are a pale reflection of Saturn's, and hardly worth the extra ten years travelling time.

Similar in structure to Jupiter, Saturn is another rapidly spinning ball of mainly liquid hydrogen, topped with a dense, banded atmosphere. Lacking the Jovian gravity, Saturn is much less dense (less dense, in fact, than water), and noticeably bloated at the equator.

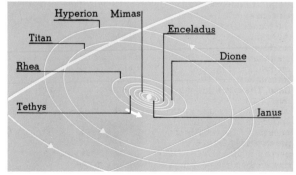

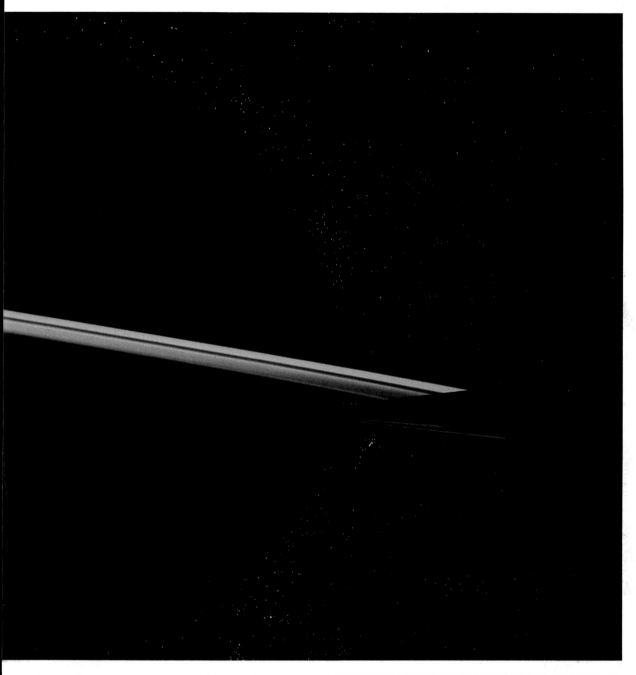

Unquestionably the most elegant spectacle among all the planets, the Rings alone make the voyage worthwhile. Highly manoeuvrable Ring Shuttles are stationed at the International Orbiting Ring Base. These are capable of negotiating the packed fields of snow-covered rocks and water-ice with remarkable accuracy. The challenging perspectives seen on a perpendicular transit of this ring system never fail to draw gasps of astonishment. The system is extremely thin — a matter of a few kilometers only — so the rings appear as a thin bright line from the inner moons (see left).

The outer fringes of the ring system lie nearly 80,000 km from the Surface of Saturn and only 22,000 km from the innermost moon, Janus. The gravitational pull of the inner satellites has helped divide the rings into separate bands.

TITAN
A planetary oasis among the Out-Planets, Titan was long known to have an atmosphere. The first robot lander in 1998 confirmed rust-coloured hydrocarbon clouds scudding across a blue sky, and ammonia ice-floes in methane slush. Volcanic activity maintains an atmosphere as dense as that of Earth, and the same 'greenhouse effect' that makes Venus so inhospitable warms Titan's surface to almost minus 50°C. The equatorial Clarke Base, built to mine hydrogen from the rich atmosphere, also has residential facilities.

The planets Uranus, Neptune and Pluto

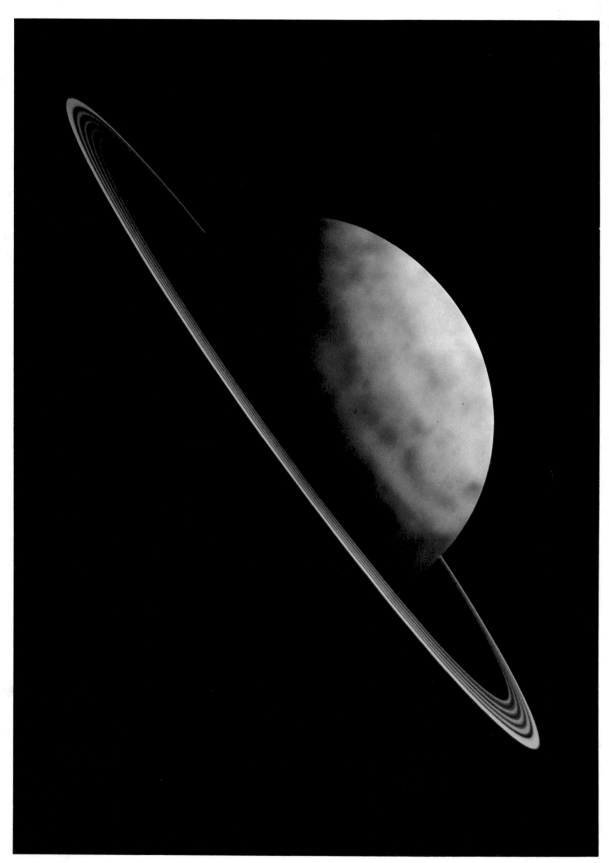

BEYOND SATURN, the interplanetary distances are much greater, and the traveller may well feel that the rewards do not merit the trouble of a very tedious journey. On the normal economic Hohmann Transfer Orbits, Uranus lies 16 years out from Earth, Neptune 30 years, while Pluto at its average distance (its orbit is quite eccentric and can sometimes lie within Neptune's) is 45 years away. Although these times can be improved by the sling-shot technique (see p. 000), normal rocketry clearly cannot meet the demands of manned travel to the fringes of the Solar System. Nuclear pulse rockets are more practical, so that consequently there is no casual traffic in these rather bleak regions.

Uranus

The ring system is the chief point of interest. First discovered in 1977 by Earth-based occultation experiments, the rings are narrow and less dense than Saturn's. The planet itself, half the size of Saturn, has a moderately thick, but surprisingly clear, atmosphere composed of hydrogen, helium, ammonia and methane at a temperature of minus 210°C.

Neptune

Very similar in size and composition to Uranus, but lacking a ring system, Neptune offers little inducement to delay the visitor. Its principal claim to fame at this time is that it has the largest moon in the Solar System — Triton, with a diameter of 6000 km, is larger than Mercury. There are no facilities here, although stores and equipment supposedly remain from the only manned visit, the 2049 expedition..

Pluto

The most remote planet in the System, Pluto is little more than a snowball of frozen methane and ammonia surrounding a water-ice core. Discrepancies in early observations created something of a mystery around Pluto in the early days, but it is now known that Christy's suggestion that the 3500 km diameter planet has a 1200 km diameter moon in close orbit is correct At this distance, from the frozen surface, the Sun appears as little more than a bright star.

Left: viewed from its second satellite, Ariel, Uranus is inclined steeply towards the Sun, as is emphasized by its five rings. At this distance, there is little light from the Sun, which is 19 times smaller than it appears from Earth.

Right: Neptune from the surface of Triton, the largest moon in the Solar System. At its furthest, more than seven billion kms from the Sun, Pluto's surface is only 40° above absolute zero. Above, its moon, Charons, sets on the horizon.

High-G events The Galaxy's predators

UNTIL RECENTLY, high-G astronomy was not thought to be of any practical interest for the astronaut, although the evolution of dwarf stars, neutron stars and black holes has powerful cosmological implications. The gravitational and radiation forces involved would make travel in their vicinity hazardous in the extreme, but, as the nearest neutron star is many light years away, it was always thought that high-G events would become navigational hazards only in the far future.

Now, however, we know differently (or at least we know enough to be less dogmatic). In 2054 the latest High-Energy Astronomy Observatory, HEAO-L, eighth in the 91 year-old series, showed a phen-

omenon that was at first difficult to explain — an intense source of X- and gamma rays, of star-like proportions, that was moving too fast for any known stellar body. The only explanation that seemed to fit was that here was a black hole *within* the Solar System! Linking the Far-Side Optical Telescope with the Goldstone tracking computer, the optical observation of this mysterious object as it occulted star HDE 217095 was an astronomical tour-de-force. The unmistakable distortion of HDE 217095 as the black hole passed in front of it, acting as a gravitational lens, proved the case. Against all the odds, an 'unanchored' black hole has come within

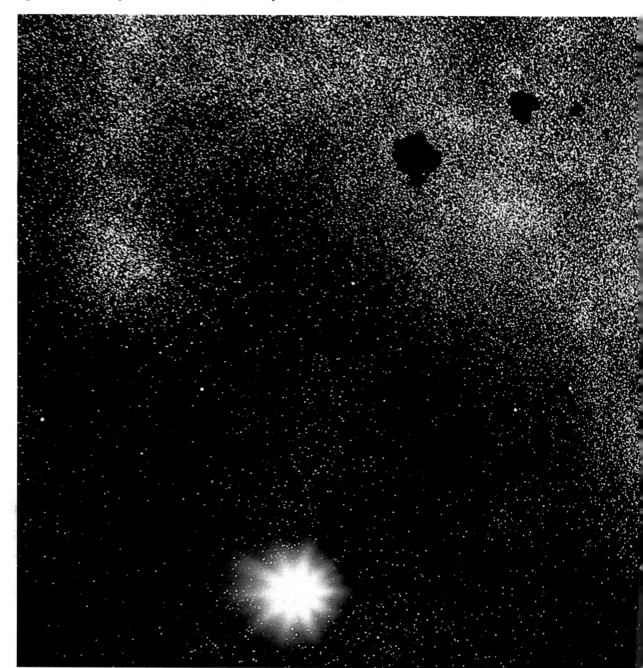

hailing distance! It is therefore all the more regrettable that a gyroscope failure on the HEAO-L caused loss of contact shortly afterwards. The best estimates, however, show that the black hole was well outside the orbit of Pluto and moving away from the Solar System.

There would appear, therefore, to be no immediate danger to space travel from black holes, but the very fact that one has been discovered so close has caused a certain amount of soul-searching among astrophysicists who have predicted that there are few black holes which move freely. As High-Energy observational instruments are not normally carried on board spacecraft — at least not those powerful enough to allow accurate focusing — the only means of detection prior to total destruction are unexplained discrepancies in the spacecraft's trajectory and a source of high radiation that appears to have no physical location. The computed effects of black holes, tidal and otherwise, are largely academic, as a spacecraft venturing too close would be rapidly stretched into molecular dust. Fortunately, the likelihood of such an encounter is extremely small.

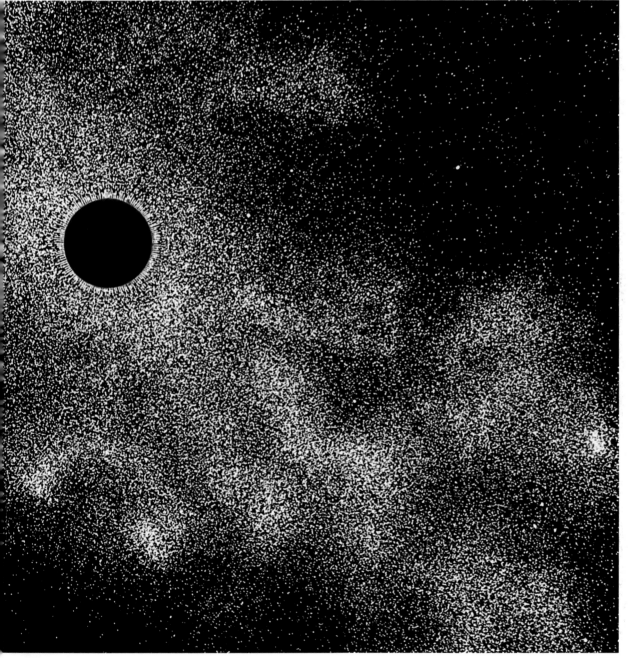

Silhouetted against an arm of the Milky Way, the black hole bends the light from distant stars.

Chronology of Space

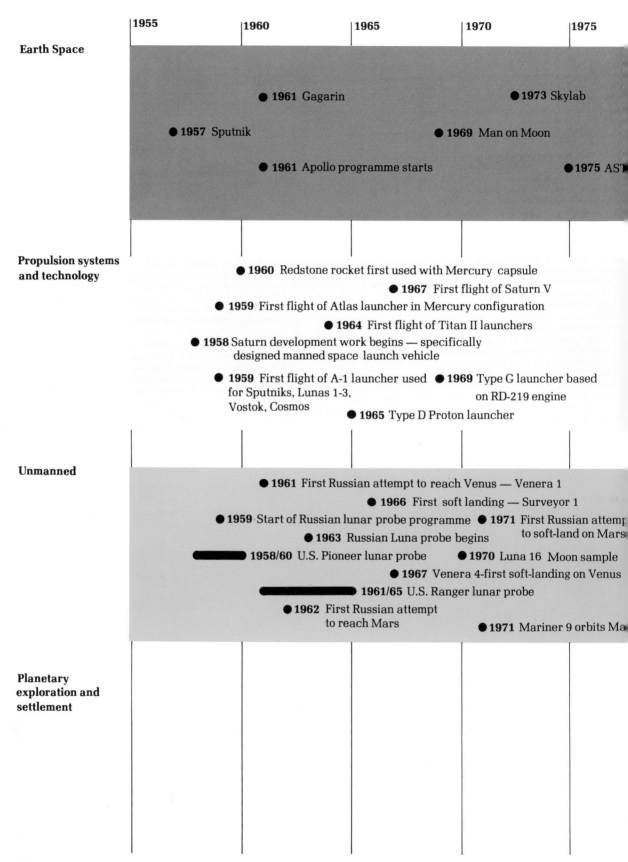

| | 1955 | 1960 | 1965 | 1970 | 1975 |

Earth Space

● **1961** Gagarin   ● **1973** Skylab

● **1957** Sputnik   ● **1969** Man on Moon

● **1961** Apollo programme starts   ● **1975** AST

Propulsion systems and technology

● **1960** Redstone rocket first used with Mercury capsule

● **1967** First flight of Saturn V

● **1959** First flight of Atlas launcher in Mercury configuration

● **1964** First flight of Titan II launchers

● **1958** Saturn development work begins — specifically designed manned space launch vehicle

● **1959** First flight of A-1 launcher used for Sputniks, Lunas 1-3, Vostok, Cosmos   ● **1969** Type G launcher based on RD-219 engine

● **1965** Type D Proton launcher

Unmanned

● **1961** First Russian attempt to reach Venus — Venera 1

● **1966** First soft landing — Surveyor 1

● **1959** Start of Russian lunar probe programme   ● **1971** First Russian attempt to soft-land on Mars

● **1963** Russian Luna probe begins

▬▬▬ **1958/60** U.S. Pioneer lunar probe   ● **1970** Luna 16 Moon sample

● **1967** Venera 4-first soft-landing on Venus

▬▬▬ **1961/65** U.S. Ranger lunar probe

● **1962** First Russian attempt to reach Mars

● **1971** Mariner 9 orbits Ma

Planetary exploration and settlement

1980   1985   1990   1995   2000

● **1996** Large space station building
        programme starts

● **1988** First manned space station

● **1994** First Moon base (Clavius)

First operational
load-carrying ion rocket **1993** ●

● **1994** First nuclear thermal rocket (→First manned
        Mars) (→First manned Saturn/Uranus) (→First
        manned Jupiter)   ● **2000** First solar power
                                              satellite

● **1980** First shuttle

● **1990** First HLLV (→First manned space station)
        (→First manned Mars expedition) (→First Moon base)

● **1992** — OTV tugs start (→First Moon base)

● **1991** Beam building machines in use   ● **2001** Purpose-built
(→ large space station construction starts)   lunar lander
                                                   (→First lunar
● **1992** First solar sail   colony)

● **1979** First detailed pictures of inner Jupiter moons — Voyager 1

                                                    1993-1998 ABM satellite 'war'

● **1992** Multi-lander probes launched to Ganymede,
        Callisto, and Io Titan

● **1979** First Saturn fly-by

                        ● **1998** Venus long-life
                                landing probe

Chronology of Space

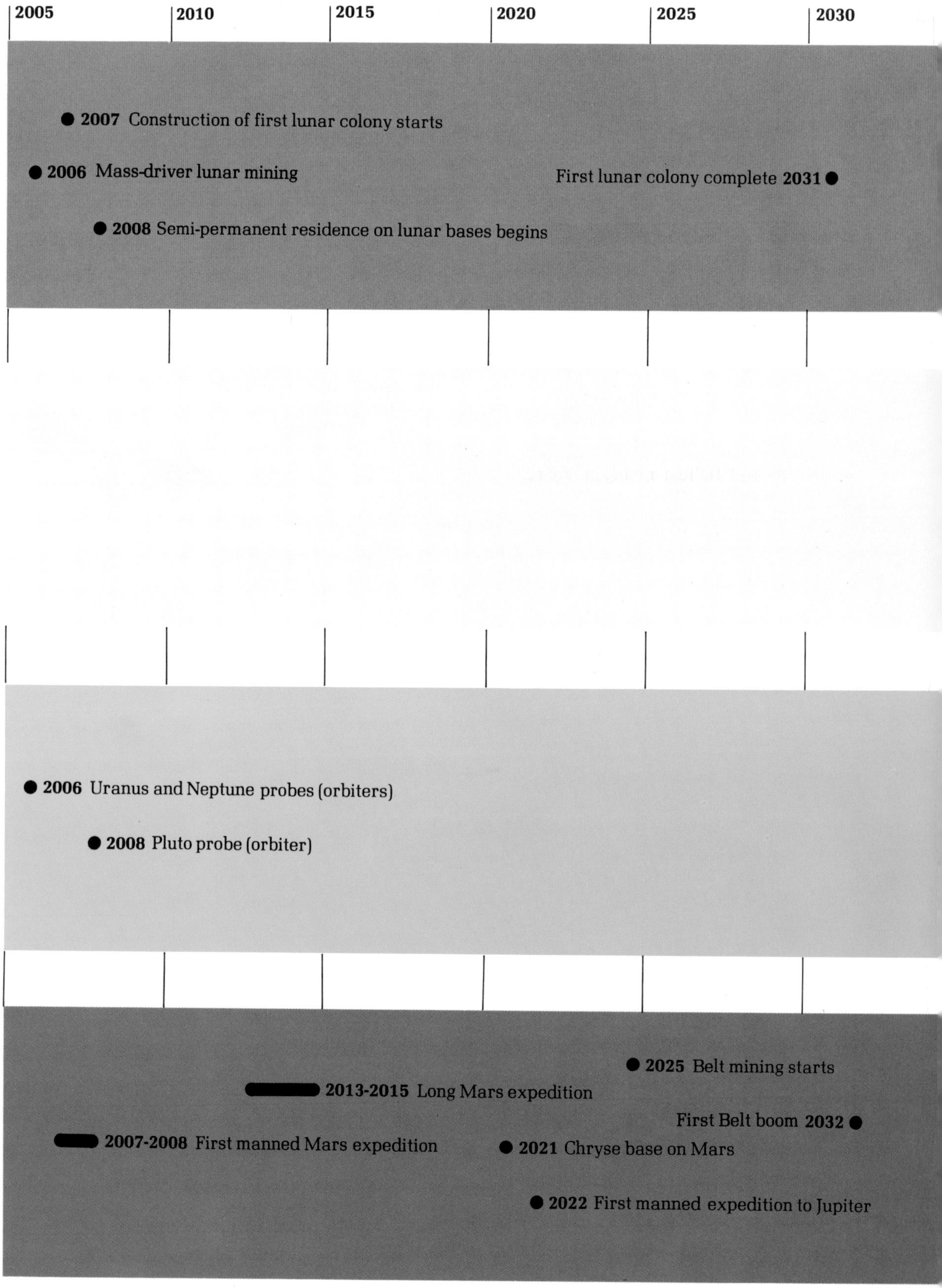

| 2005 | 2010 | 2015 | 2020 | 2025 | 2030 |

● **2007** Construction of first lunar colony starts

● **2006** Mass-driver lunar mining   First lunar colony complete **2031** ●

● **2008** Semi-permanent residence on lunar bases begins

● **2006** Uranus and Neptune probes (orbiters)

● **2008** Pluto probe (orbiter)

● **2025** Belt mining starts

▬▬ **2013-2015** Long Mars expedition

First Belt boom **2032** ●

▬▬ **2007-2008** First manned Mars expedition   ● **2021** Chryse base on Mars

● **2022** First manned expedition to Jupiter

2035	2040	2045	2050	2055	2060

2037-2050 Peak of Lagrange colony building

● **2040** Helium mining in Jupiter

● **2042** First nuclear pulse rocket
(→Daedalus construction starts)

● **2056** onwards: Ramjet development

● **2049** Daedalus departs

● **2042** Daedalus construction starts

● **2045** Belt Charter

● **2051** First Mercury expedition

● **2060** Mars terraforming tests

● **2055** Clarke base on Titan

● **2038** Callisto base
(→Helium mining in Jupiter)

● **2048** First manned Saturn expedition

● **2055** Petroleum boom in Belt

● **2046** Loss of Ganymede Star

● **2052** Venus terraforming tests

2049 ● First manned expeditions
to Uranus and Neptune

● **2057** Saturn's International
Orbiting Ring Base

Glossary

accelerometer - device for measuring rate of change of velocity (acceleration), usually in either direction along a particular line or axis, by means of a suspended inertial mass.

aerosol - very small particles of dust, or droplets of liquid, suspended in the atmosphere.

angular momentum - a measure of a body's tendency to continue rotating at a particular rate (or at rest) around a particular axis; obtained by multiplying the moment of inertia of a body by its angular speed.

aphelion - the farthest point out in an orbit around the Sun; opposite of perihelion.

apogee - the farthest point out from Earth in an elliptical orbit. To enlarge or circularize the orbit, a spacecraft's thruster is turned on at apogee to give the craft an 'apogee kick'.

asteroid belt - a solar orbit zone or doughnut-like ring located between 2.1 and 3.5 astronomical units in which are found thousands of asteroids (also called minor planets or planetoids) of irregular shapes and diameters from a fraction of a kilometer to 800 km.

astronomical unit - a measure of Solar-System distance equal to the average distance between Earth and Sun; its value is approximately 149,000,000 km (93,200,000 miles).

atmosphere - can have different meanings. (1) The Earth's atmosphere is 80 percent nitrogen and 20 percent oxygen. The density and pressure decrease with altitude and are barely detectable at 200 kilometers (see **drag**). (2) **Cabin** atmosphere is normally either ordinary air at sea-level pressure or almost pure oxygen at one third that pressure. (3) Atmosphere is also a common unit of gas pressure.

atmospheric probe - scientific device, usually carried by a spacecraft, for determining the pressure, composition and temperature of a planet's atmosphere at different altitudes.

attitude - the direction toward which a spacecraft is pointing, usually defined by the directions of its X-, Y- and Z- axes relative to the stars.

attitude control - process of maintaining or changing a spacecraft's orientation in space (usually with gas jets) so that solar panels and other instruments can be pointed at target bodies.

ballistic trajectory - unpowered flight similar to a bullet's trajectory, governed by gravity and by the body's previously acquired velocity.

booster rocket - the large reaction motor used to launch a spacecraft.

bow shock wave - the interface formed where the electrically-charged solar wind encounters an obstacle in space such as the atmosphere or magnetic field of a plane.

celestial mechanics - dynamic relationships existing among bodies of the solar system; description of the relative motions of celestial bodies under the influence of their mutual gravitational forces.

charged particle detector - a device which counts and/or measures the energy of electrically charged particles (electrons, protons, alpha particles, larger ions) in space.

charged particle telescope - a group of charged particle detectors used to measure the direction of particle passage, as well as for counting the number of such particles.

circuit - communications link between manned spacecraft and ground stations. Some circuits are reserved for voice, television, data telemetry or computer.

circularize - to change an elliptical orbit into a circular one, usually by 'apogee kicks'.

Command Module (CM) - a component of spacecraft, attached to the Service Module (SM) until re-entry into the Earth's atmosphere, when the SM is jettisoned.

configuration - a spacecraft assembly at a particular point in its voyage. One spacecraft may have several configurations over the course of its mission as stages are jettisoned.

corona - the glowing outer reaches of the Sun's luminous, active, gaseous envelope, visible during a total eclipse; includes towering prominences and shades off into the invisible, tenuous solar wind (plasma) which streams out into the solar system.

cosmic dust - fine microscopic particles adrift in space; sometimes called micrometeorites.

cosmic rays - atoms with all the electrons removed (see ion) moving at relativistic speeds from galactic sources.

docking - sealing two spacecraft together in orbit with latches and sealing rings so that two hatches can be opened between them without losing cabin atmosphere. The docking target is used by the crews to align the spacecraft so that latches fit into hooks.

docking module - a special component added to spacecraft so that they can be docked with other craft.

dosimeter - instrument which measures radiation doses.

drag - atmospheric resistance to the orbital motion of a spacecraft. The effect of drag is to lower the orbit. Above 200 km, the altitude decreases very slowly. Below 150 km, the orbit 'decays' rapidly.

Earth - third planet, averaging 149,598,000 km from the Sun. Very nearly a sphere of 6378 km radius, 6×10^{24} km mass. The Earth is accompanied by the Moon, about one fourth its size and 384,405 km distant.

eccentricity - a measure of the ovalness of an orbit. When $e = 0$, the orbit is a circle; when $e = 0.9$ it is a long, thin ellipse.

eclipse - covering a bright object with a dark one. In a normal solar eclipse, the Sun is covered by the Moon. Apollo covered the Sun for Soyuz.

ecliptic - the plane of motion defined by Earth's orbit around the Sun in space.

electromagnetic radiation - energy transmitted through space in any of the following forms: radio waves, infrared radiation, radiant heat, visible light, ultraviolet rays, x-rays or gamma rays.

ellipse - a smooth, oval curve, accurately fitted by the orbit of a satellite around a much larger mass.

encounter - a close flyby or rendezvous of a spacecraft with a target body.

energy - the capability of doing work. Kinetic energy is the energy of motion and is equal to $\frac{1}{2}mv^2$. Potential energy depends on position and is larger the farther mass m is from Earth.

entry probe - see atmospheric probe.

escape velocity - the speed necessary to escape from Earth's gravity. It is smaller the farther a spacecraft is from Earth.

exobiology - the study of extra-terrestrial environments for living organisms, the recognition of evidence for possible existence of life in these environments and the study of any extra-terrestrial life that may be found.

flyby - space mission in which instrumented vehicle passes a planet without going into orbit, entering its atmosphere or landing on its surface.

force (F) - a push or pull on a mass m that produces an acceleration a; $F = ma$.

free fall - when a spacecraft is moving solely under the force of gravity (no drag, no thrust).

Galilean satellites - the four large satellites of Jupiter: Callisto, Ganymede, Europa and Io; discovered by Galileo in 1609.

gamma rays - very high energy photons of wavelength shorter than x-rays and energy higher than x-rays. They are produced by nuclear reactions and other processes in distant regions of space.

geomagnetic storm - sudden worldwide fluctuations in Earth's magnetic field, associated with solar flare-generated shock waves which propagate from the Sun to Earth.

geosynchronous orbit - an orbit that is synchronized with the Earth's rotation. A satellite 35,900 km above the equator with a period of 24 hours would be in a geosynchronous orbit; it would always be above the same point on Earth.

gravity anomaly - a region where gravity is lower or higher than expected if the Earth's crust is considered to have uniform density.

gravity assist - change in a spacecraft's velocity and direction achieved by calculated flyby through a planet's gravitational field without use of supplementary propulsive energy.

Greenwich mean time (GMT) - the time of an event, from 0 at midnight to 12 hours at noon to 24 hours at midnight, as measured at 0° longitude (Greenwich, near London, England).

ground elapsed time - the time from launch.

hatch - a door in the pressure hull of a spacecraft. The hatch is sealed tightly to prevent the cabin atmosphere from escaping to the outside vacuum.

heliosphere - the region in the Solar System occupied by the Sun's corona, including the solar wind, which is known to extend beyond Earth.

imaging photopolarimeter - sensor for measuring brightness and polarization of light; usually mounted to scan target so that readings can be assembled into a picture.

infrared radiometer - instrument which measures the temperature of an object from the intensity of radiated heat.

inner planet - Mercury, Venus, Earth and Mars (see terrestrial planets).

Inner System - the inner planets considered together.

interplanetary medium - the environment of charged particles and associated magnetic fields existing in the Solar System outside the regions affected by the atmosphere, ionosphere or radiation belts that envelop individual planets.

interstellar medium - the environment of charged particles and dust that exists in the Milky Way Galazy throughout the region between the stars.

ion - an atom with one or more electrons removed or, more rarely, added. Cosmic-ray ions have all electrons removed and ionize other atoms as they pass them at high speed.

ionosphere - electrically charged upper layer in an atmosphere which is ionized by the Sun's ultraviolet and x-radiations.

jettison - to discard. When the fuel in a booster rocket is used up, the now-useless booster is disconnected from the spacecraft and jettisoned (allowed to fall back to Earth).

Kepler's Third Law - the law which states that T^2 is proportional to A^3, where T is the period and A measures the orbit size. Based on early observations of planets, the law also applies to satellites of the Earth and is explained by Newton's Laws.

lander - spacecraft or mission which lands on another celestial body.

launch configuration - the combination of boosters, spacecraft and launch system that must be lifted off the ground at launch.

LOX - liquid oxygen at temperature 90K or —183°C, used with kerosene fuel as a propellant in booster rockets.

magnetic field - the strength of the magnetic force on a unit magnetic pole in a region of space affected by magnets or electric currents.

magnetic lines of force - theoretical lines giving the direction of the force on a test magnetic pole at any place in a magnetic field.

magnetometer - device for measuring the strengths of magnetic fields.

magnetopause - the outer boundary of the magneto-sphere.

magnetosheath - the region of disturbed solar wind which lies between the bow shock wave and the magnetopause.

magnetosphere - the region (not actually spherical) within which the magnetic field of a planet is confined by the solar wind.

mascon - area of mass concentration or high density within a planetary body, usually near the surface.

mass spectrometer - device that separates a stream of charged particles into a spectrum according to the masses of the particles; used for measuring the atomic masses.

micrometer - one millionth of a meter, formerly called a micron.

Milky Way Galaxy - a disc-shaped group of more than 100 billion stars, including our Sun.

mission control centre - the operational headquarters of a space mission.

momentum - mass times velocity, referring to motion in a straight line. *Angular momentum* refers to rotation and to motion around orbits. It is mass times cross-velocity times distance from the axis of rotation or centre or orbit. Both are conserved.

muli-stage launch - a launch that uses several stages to boost the payload into orbit. After the first-stage booster uses its fuel, it is jettisoned and the secondary booster is fired. When the second-stage fuel is gone, that booster is jettisoned, and so on. Such multi-stage launching allows very high payload velocities.

nacelle - aerodynamic housing, usually for an engine.

neutral atmosphere - that portion of an atmosphere consisting of atoms and molecules, not electrically-charged ions.

neutron star - a very high density star made of neutrons, not atoms.

newton - metric unit of force, equivalent to 0.2247 pounds. It is equal to the force that will give a mass of 1 kg an acceleration of 1 meter/sec^2.

Newton's Laws - the three laws of motion and the law of gravitation, published in 1687, explaining almost all the motions of planets and satellites with a high degree of accuracy.

nuclear power - power derived from nuclear reactions between neutrons and atoms of uranium, thorium or plutonium, which undergo fission (splitting). Such power may be used for reaction motors. The fission products are highly radioactive.

occultation - a blocking action, as when the Moon passes between a star and an observer, cutting off its light; a planet with an atmosphere gradually cuts off light or radio waves from a spacecraft and the occulting atmosphere can be studied by this means.

orbit - the path followed by a planet around the Sun or by a satellite around a planet, usually an ellipse.

orbiter - spacecraft or mission involving insertion of vehicle into orbit about another celestial body.

outer planets (or Out-planets) - Jupiter, Saturn, Uranus, Neptune and Pluto; the first four differ radically from Earth and the other inner planets.

parabolic antenna - a radio reflector used for radar, microwave and space communications; in receivers the reflecting surface reflects parallel beams to a single focal point, where the active element of the antenna is located; in transmitters the reflecting surface converts the source signal to a parallel beam.

parking orbit - a temporary orbit (usually around Earth) in which a space vehicle coasts or 'parks' between intervals of powered flight before injection into a transfer trajectory to another body in the solar system.

particles and fields - cosmic dust, plasma, other charged particles including cosmic rays and magnetic fields; the term usually refers to instruments and experiments conducted in interplanetary space.

pascal - a unit of pressure (Pa) equal to pressure of 1 newton/sq meter.

payload - the components to be put into orbit on a single-stage launch. On a multi-stage launch, the second stage is payload for the first; the third stage is payload for the second, and so on.

perigee - the point closest to Earth on an elliptical orbit around the Earth.

perihelion - in a solar orbit, that point in the ellipse closest to the Sun.

period - the time taken by a satellite to travel once around its orbit.

photopolarimeter - see imaging photopolarimeter.

plane of the ecliptic - see ecliptic.

planetismal - a large asteroid.

plasma - an ionized gas containing about equal numbers of positive ions and electrons, as in the solar wind.

plasma detector - device for measuring the amount and/or velocity and direction of solar plasma.

propellant - both the fuel and the oxidizer for a reaction motor. The propellant is ejected at high velocity to provide forward thrust.

proton - a positively charged atomic particle; one proton constitutes the nucleus of the hydrogen atom.

rad (radiation absorbed dose) - a unit of radiation damage to living organisms; equal to 10^{-5} joule absorbed per gram of tissue.

radar altimeter - device for measuring range or distance; for example, from an approaching spacecraft by timing the travel of a radar pulse down to the surface and back.

radar astronomy - study of the motion and form of other planets with powerful and precise radar equipment; differs from radio astronomy in that signals are transmitted by the observer and reflected by the object of interest.

radiation - a term used loosely to include cosmic-ray particles and high-energy protons, as well as penetrating electromagnetic waves (x-rays and gamma rays).

radioisotope thermo electric generator - a source of electrical power in which the energy liberated in radioactive decay is collected as heat and converted, usually directly by means of thermocouple action, into electricity.

radio telescope - a precise and sensitive radio receiving system using a parabolic or other highly directional antenna to locate and track radio sources in the sky.

RCS quad jets - small jets used to roll or rotate a spacecraft.

reaction - the equal but opposite push on, e.g. your hand, when you push something (Newton's Third Law). Reaction motors push gas out of the rear nozzle to get the reaction as a forward thrust.

rendezvous - a space mission in which the spacecraft is manoeuvred so as to fly alongside a target body, such as a comet or asteroid, at zero relative velocity.

rover - a roving vehicle, either manned or remotely controlled, for planetary or lunar surface exploration.

S-band - a radio frequency band between 1550 and 5200 megahertz (VHF and UHF television signals have frequencies between 50 and 900 megahertz).

sealing rings - mechanical devices designed to fit tightly when two spacecraft are docked so that cabin atmosphere will not leak out.

Service Module (SM) - the large part of Apollo spacecraft that contains the main thruster, tanks, radio equipment, and other support equipment. It is attached to the CM until just before the Cm re-enters the Earth's atmosphere.

shield - an absorbing material that prevents the passage of radiation or reduces its intensity.

solar electric propulsion - a relatively low-thrust, long-continuing method of propulsion in which stored matter, called reaction mass, is given very high velocity and jetted out by means of electrical energy generated from solar panels.

solar panel - a wing-like set of cells that convert sunlight to electric power.

solar wind - plasma blown constantly at supersonic speed (20km/sec) out of the Sun in all directions; consists of electrons, protons and alpha particles (hydrogen and helium atomic nuclei, both positively charged), and some heavier ions.

specific impulse - a measure of the power of a propellant.

spectral survey - measurements of the wavelengths or energies of radiation emitted by a given source or from all accessible regions of the celestial sphere.

stage - one part of the launch sequence; see multi-stage launch.

telemetry - the automatic transmission of data to ground receivers.

telescope - an instrument for measuring the direction

of incoming rays.

terminal guidance - navigation of a spacecraft, usually during its approach to a planet, by observing the angular position and motion, and perhaps also the apparent size of the target body.

terminator - the shadow line around a planet or satellite which separates the sunlit from the shaded side.

terrestrial planets - Mercury, Venus, Earth and Mars; the planets in the Inner Solar System which fall in a class with our own as to size and density.

thermocouple action - the process whereby, in an electrical circuit made of two dissimilar metals or semiconductors, with one junction heated and the other cold, an electric current is generated (see radioisotope generator).

thrust - the forward force provided by a reaction motor.

time line - the planned schedule for astronauts on a space mission.

time zone - a region using the same time of day. There are 24 time zones around the world, each about 15° in longitude. In the United States, they are called eastern, central, mountain and Pacific standard time, each one hour different from the zone on either side.

torque - a twist provided by two offset forces on a body.

transfer trajectory - that part of a spacecraft's travel in space between, usually, Earth and a target body; usually unpowered or purely ballistic.

trapped radiation - charged particles of moderately high energy, trapped by a planet's magnetic field.

ultraviolet - invisible light of wavelengths less than 4000 angstroms (400 nanometers), shorter than those of visible light.

ultraviolet airglow spectrometer - an ultraviolet spectrometer designed especially to define the faint fluorescent glow of ultraviolet light in a planetary atmosphere in order to study the gases or ions and energy sources which produce it.

ultraviolet spectrometer - an optical instrument for analyzing the intensity of ultraviolet light at various wavelengths.

Van Allen Belt - a doughnut-shaped region around the Earth from about 320 to 32,400 km (200 to 20,000 miles) above the magnetic equator, where high-speed protons and electrons oscillate north-south in the Earth's magnetic field.

vector - a directed quantity, like velocity, force, acceleration (v, F, a).

velocity - change of position per unit time, in meters per second.

weight - the downward force on a mass at the Earth's surface. The force on 1 kg is 9.8 newtons.

weightlessness - the condition of free fall or zero-G, in which objects in a spacecraft are weightless.

X-band - a radio frequency band from 5200 to 10,900 megahertz, designated originally for high-frequency radar; now used for space communications.

X_v, Y_v, Z_v, - spacecraft (vehicle) axes, with X_v directed forward (away from the thruster nozzle), Y_v to one side, and Z_v 'up'.

zero-G - the condition of free fall and weightlessness.

zodiacal light - a faint glow around the general region of the plane of the Solar System; thought to be sunlight reflecting from particles of cosmic dust found mostly in this plane.

Picture Credits

Michael Freeman: Facing title page, **4, 6, 8, 12, 14** (top), **16, 17, 18, 19, 32** (top), **36, 37** (top), **38** (top), **39, 40-1** (across centre), **43, 44/5, 47** (bottom), **50, 52** (top left, top right), **56/7, 59** (top, middle, bottom right), **65**(3), **66, 67** (left and bottom right), **69** (bottom 2), **75, 76/7, 84, 86** (top), **94, 95, 100, 101** (except top), **103, 114/5** (fuel cell & solar panel), **116, 118, 126, 127, 131** (bottom), **136, 137, 144, 151** (top), **160, 164** (bottom), **166/7, 168, 180/1, 182/3, 190/1, 192/3, 194/5,**
Novosti press agency **62, 64** (top 2 and bottom)
Fotokhronika Tass **63** (bottom)
British Interplanetary Society **96/7**
Royal Astronomical Society **155** (bottom) **171** (top & middle)
Biophoto associates, Dr G. F. Leedale **165** (bottom left)
United Nations, M. Grant **48/9**
All other photographs were supplied by NASA, to whom the publishers express their gratitude for their kind assistance. They are not, of course, responsible for any shortcomings in the book.

CONVERSION TABLE

Linear measure
1 millimeter = 0.03937 inches
10 millimeters = 1 centimeter = 0.3937 inches
100 centimeters = 1 meter = 39.37 inches or 3.2808 feet
1000 meters = 1 kilometer = 0.621 miles or 3280.8 feet

Square measure
1 square millimeter = 0.00155 square inches
100 square millimeters = 1 square centimeter = 0.15499 square inches
10,000 square centimeters = 1 square meter = 1549.9 square inches or 1.196 square yards
1,000,000 square meters = 1 square kilometer = 0.386 square miles or 247.1 acres

Volume measure
1000 cubic millimeters = 1 cubic centimeter = 0.06102 cubic inches

1,000,000 cubic centimeters = 1 cubic meter = 35.314 cubic feet or 1.308 cubic yards

Capacity measure
10 milliliters = 1 centiliter = 0.338 fluid ounces
100 centiliters = 1 liter = 1.0567 liquid quarts or 0.9081 dry quarts
10 liters = 1 decaliter = 2.64 gallons or 0.284 bushels

Weights
10 milligrams = 1 centigram = 0.1543 grains or 0.000353 ounces
100 centigrams = 1 gram = 15.432 grains or 0.035274 ounces
1000 grams = 1 kilogram = 2.2046 pounds
1000 kilograms = 1 metric ton (1 tonne) = 2204.6 pounds

Index

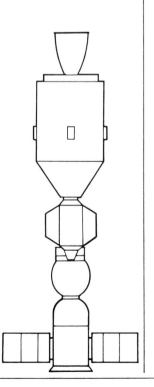

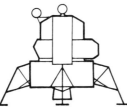

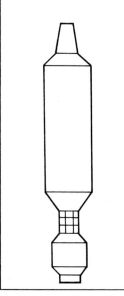

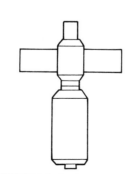

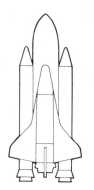

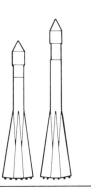